Basic Building Craft Science

Basic Building Craft Science

Alf Fulcher   Brian Rhodes   Bill Stewart   Derick Tickle

John Windsor   John Taylor

OXFORD

BSP PROFESSIONAL BOOKS

LONDON EDINBURGH BOSTON

MELBOURNE PARIS BERLIN VIENNA

First published in Great Britain by
   Granada Publishing Limited 1981
   ISBN 0–246–11223–9
   ISBN 0–246–11265–4 (Pbk)
Reprinted 1982
Reissued by BSP Professional Books 1990

British Library
Cataloguing in Publication Data

Basic building craft science.
   1. Science
   I. Fulcher, Alf
   500

   ISBN 0-632-02418-6

BSP Professional Books
A division of Blackwell Scientific
   Publications Ltd
Editorial Offices:
Osney Mead, Oxford OX2 0EL
   (Orders: Tel. 0865 240201)
25 John Street, London WC1N 2BL
23 Ainslie Place, Edinburgh EH3 6AJ
 3 Cambridge Center, Suite 208,
   Cambridge, MA 02142, USA
54 University Street, Carlton, Victoria 3053,
   Australia

Set by V & M Graphics Ltd, Aylesbury, Bucks
Printed in Great Britain at the Alden Press, Oxford

CONTENTS

PREFACE

This book has been written for the building craft student who has had very little basic science education before entering college. It is written as a simple reference book both for students, and as a basis for lesson preparation for teachers.

We have attempted to isolate the essential scientific principles from the City and Guilds of London Institutes syllabuses which can be directly related to building work, and then to explain them in non-scientific language. To achieve this we have had to take the occasional liberty with accepted scientific methods and teachings. When it has not been possible to explain a process or action without using technical language or jargon we have offered it as a fact without explanation.

We trust that it remains a non-science science book and not a nonsense science book. We hope that if the facts are not always as simple as they may appear here, at no time are they inaccurate. We have tried to write it in such simple language that at no time can our meaning be misunderstood.

Because science is an enquiring subject we have adopted a question and answer approach. Some questions have had to be contrived to bring in what we consider to be an important aspect of craft knowledge or to explain the principle behind another question.

In trying to avoid scientific terms some answers have had to be contrived as well. An example of this concerns acids and alkalis. To describe alkalis required a discourse on non-metallic elements, combinations with water, and the giving off of hydroxyl ions. This, we felt, was too advanced for the level of this book so we approached it by asking how alkalis can be detected and what effect they have on building materials.

Some critics may state that to simplify the subject to this level jeopardises the chances of students who wish to advance to TEC courses. Our answer is that only a very small proportion of craft students take this route. Those who have the ability to qualify for TEC would probably have a sound scientific knowledge before starting on the craft course, and would make it in spite of us. We would add that this simple approach to building science could possibly help TEC students who are having trouble with science.

Undoubtedly the simplicity of the text will be of value to secondary school students, particularly those who are hoping to go into the building industry. It may not be a text book for G.C.E. or C.S.E. but many students could be helped by it.

In no way are we suggesting that craft science should be taught only at this level. It is a subject which is tackled by colleges in a variety of ways, with a mixture of success. What we are suggesting is that a student who achieves our level of competency is possibly a better craftsman than one who has never understood the purer science that has been directed at him. Undoubtedly it is a better level for those who have attended colleges where science has been ignored. Also it may help those teachers who have either thought the subject irrelevant or difficult because of their own poor scientific background.

Most answers have been prepared in three parts. The first part answers the question in a short, easy to remember fashion. This may be sufficient for some students or occasions. The second part explains the scientific principle or action related to the question. The last part gives examples where the principles apply, either to building or everyday life.

In each answer key words or phrases have been picked out and shown in capitals. These are standard scientific terms. Every effort has been made to explain them, or relate them to building. But for the more ambitious or enquiring student each key term has its brief scientific definition in the glossary at the back of the book.

We have used illustrations for two purposes. First, to make the book more attractive. Secondly to help describe a principle. We have avoided using illustrations to answer questions at the expense of text. Language is the principal method of communication expected of students, par-ticularly for examinations, so we have included only matters which can be defined in words.

The book contains no laboratory experiments. There are numerous excellent books on craft science which emphasise experiments. We are not in competition with these and recommend readers to them if they require experiments as an aid to learning.

In our conceit we hope that City and Guilds question writers might read our book with interest. It is a unique approach to written craft science and, if it fills the need of craft students, questions of this sort might be set to those studying for certificates.

This is not a definitive book on craft science. No doubt some may argue with the selection of certain topics and would feel that some other areas should have been covered. In which case we will welcome criticism and suggestions from teachers and students, and if the opportunity arises we will include the omissions and omit the irrelevances in any revised editions.

ACKNOWLEDGEMENTS

To list all the publications that we have referred to in the writing of this book would require a second volume. If any authors think that they recognise their writings amongst our words may we take this opportunity to say thank you.

Other people were not quite so lucky and were approached face to face. To these people we acknowledge our thanks formally. Most of these were our college science and craft colleagues whose advice has been invaluable, often encouraging, and always gratefully received.

Those whom we approached outside of the colleges and to whom we would like to express our thanks are Hilary Fairless who read some of our first drafts and encouraged us, and Doug Marshall who tried to make some of the paint technology simple enough for us to understand.

Particular thanks must go to Jasmine, Julie, Lily, Margaret and Sue who missed out on many things while we struggled with the text. We offer a special thanks to Lily Stewart who typed the complete manuscript . . . and not only once.

BASIC SCIENCE

1.1 What are solids, liquids and gases?

They are the three forms in which substances can exist. Solid, liquid and gas are known as the three STATES OF MATTER.

Although there are millions of substances they exist only in either solid, liquid or gas forms. The difference between each state is the amount of freedom of movement of the MOLECULES. Also the strength of attraction or bonds between them.

SOLIDS (Fig. 1.1)
In solid materials the molecules do not move much. They shake together, closely bonded in a fixed position. The strength of bonding is called COHESION. It is the cohesion of molecules that keeps the fixed shape of solids.

LIQUIDS
The molecules of liquids also shake or vibrate. But because the bonds between them are weaker, the molecules are able to move and change position in all directions. This movement prevents liquids from forming a fixed shape and allows them to flow. They have less cohesion than solids.

GASES
The molecules of gases move so much that the bonds are always broken. The molecules spread out in all directions leaving large amounts of space between them. Molecules of gas have no cohesion.

1.2 Are different substances made of totally different things?

Not necessarily. Although there are millions of different substances, all these are made from a selection of about one hundred basic units called ELEMENTS.

All substances, whether naturally occurring or man-made, either *are* elements or are *formed from* a selection of these elements combined in many different ways.

An element is made of one kind of ATOM only and cannot be broken down into anything simpler. Elements are the building blocks of all other substances.

Some elements can be chemically combined with each other in huge numbers of combinations to make millions of different substances called COMPOUNDS. Many common substances are compounds. For example, chalk is the combination of the elements calcium, carbon and oxygen. Sand is a compound also. It is formed from the element silicon and the element oxygen. See table 1, 'Elements in substances'.

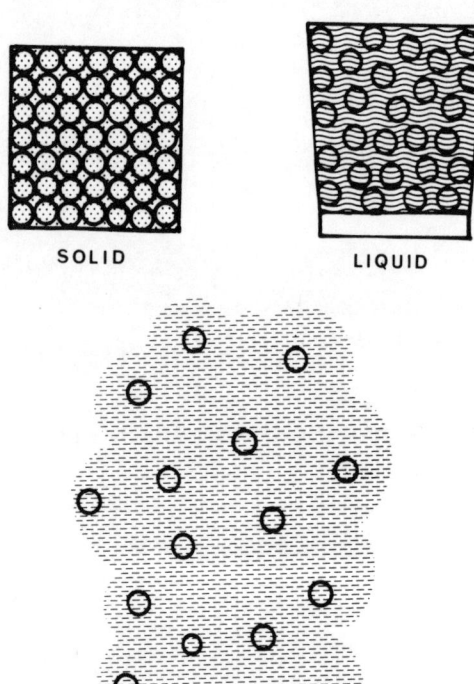

SOLID   LIQUID

Fig. 1.1   G A S

Sometimes both elements and compounds can be mixed together without any chemical action taking place. Substances made in this way are called MIXTURES. Putty is a mixture made from two compounds, vegetable oil and chalk. Brass is a mixture of two metallic elements, copper and zinc.

Elements can be either metals or non-metals. Generally metallic elements are those which can conduct electricity. There are more metal elements than non-metal elements. Gold, silver, iron and lead are elements which are metals.

Each element can be shown by a CHEMICAL SYMBOL. For example, the chemical symbol for iron is Fe. See table 2, 'Common elements and their symbols'.

Table 1 **Elements in substances**

Substance	Elements
Water	Hydrogen and oxygen
Alcohol	Carbon, hydrogen and oxygen
Sugar	Carbon, hydrogen and oxygen
Butane	Carbon and hydrogen
Ammonia	Nitrogen and hydrogen
Salt	Sodium and chlorine
Washing soda crystals	Sodium, carbon, oxygen and hydrogen

Table 2 **Common elements and their symbols**

Metals

Aluminium	Al	Mercury	Hg
Chromium	Cr	Potassium	K
Copper	Cu	Silver	Ag
Gold	Au	Sodium	Na
Iron	Fe	Tin	Sn
Lead	Pb	Titanium	Ti
Calcium	Ca	Tungsten	W
Magnesium	Mg	Zinc	Zn
Manganese	Mn		

Non-metals

Carbon	C	Oxygen	O
Chlorine	Cl	Phosphorus	P
Hydrogen	H	Silicon	Si
Nitrogen	N	Sulphur	S

1.3 Are atoms in everything?

Yes. All substances are made of very tiny particles called ATOMS.

Atoms are so small that even very little amounts of substances contain enormous numbers of atoms. The flat end of a pin, a drop of water, and a whiff of butane gas are all made of billions of atoms. A drop of water contains about 100 000 000 000 000 000 000 atoms.

Atoms have electrical forces which attract them to other atoms. These forces or bonds make atoms join with other atoms to form groups called MOLECULES. When two atoms of hydrogen bond with one atom of oxygen one molecule of water is formed (Fig. 1.3). One molecule of water is the smallest amount of water that can exist.

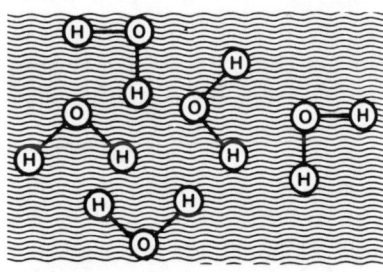

two atoms of hydrogen + one atom of oxygen = one molecule of water

Fig. 1.3

A substance changes if the number and arrangement of atoms in the molecule is altered. If more atoms are added to the molecule a new substance is formed.

If one more atom of oxygen is added to the molecule of water, one molecule of hydrogen

15

peroxide is produced. This chemical is very different from water. Hydrogen peroxide is a powerful bleach and has other uses in medicine and rocket fuels. It is extremely dangerous to drink.

In a propane molecule three carbon atoms are bonded with eight hydrogen atoms. A butane molecule has four carbon atoms bonded with ten hydrogen atoms.

The number of atoms in different molecules can vary tremendously. A sugar (sucrose) molecule contains a total of forty-five atoms. A molecule of protein may contain many thousands of atoms linked together.

1.4 What does a chemical formula mean?

It is a kind of shorthand which shows the ELEMENTS that make up a substance.

A MOLECULE is the smallest amount of a chemical COMPOUND that can exist. A chemical formula describes one molecule of a compound. It shows the ELEMENTS, and the number of ATOMS of each element in the molecule. For example, water contains two atoms of hydrogen and one atom of oxygen. Quick lime contains one atom of calcium and one atom of oxygen. Slaked lime is quick lime which has joined with water, therefore it contains one atom of calcium and two atoms each of hydrogen and oxygen.

Instead of using the elements name in a formula, a symbol is used. The symbol consists of the first letter, or the first letter and another letter, of the element's name. For example, oxygen is represented by O, and calcium by Ca. English names are not always used, so some symbols do not seem to fit the elements name. For example, the Latin name for lead was plumbum, so lead has the symbol Pb.

The number of atoms of each element is shown in the formula by a figure after the symbol except when there is only one atom. For example, Ca = one atom of calcium, and O_2 = two atoms of oxygen. If symbols are shown in brackets, such as $(OH)_2$, this means it contains two atoms of oxygen and two atoms of hydrogen.

The formula for water is H_2O, and for quick lime is CaO. When these two are joined to form slaked lime the formula is $Ca(OH)_2$.

As well as a chemical formula, substances have chemical names. The name is based on the elements that make the substance. The use of formula and name makes sure that substances are accurately described. See table 3, 'Common substances'.

Table 3   **Common substances**

Common name	Formula	Chemical name
Rust	$Fe_2O_3 + H_2O$	Ferric oxide
Quicklime	CaO	Calcium oxide
Caustic soda	NaOH	Sodium hydroxide
Red lead	Pb_3O_4	Lead oxide
Chalk	$CaCO_3$	Calcium carbonate

1.5 What is density?

Density is the amount of matter in a particular volume of a substance.

DENSITY is a property that all substances have. It is affected not only by the kinds of molecules in the substance but also by the spaces between them.

For example, one cubic metre of air in a room weighs about 1 kg. The density of room air is roughly 1 kg per cubic metre. In other words, it is its weight per volume. One cubic metre of water weighs 1000 kg. Therefore, the density of water is 1000 kg per cubic metre.

Surprisingly enough the molecules of the gases in air are heavier than water molecules. But because the water molecules are very much closer together the same amount of water is about 1000 times heavier than the same amount of air.

One cubic metre of lead weighs over 11 000 kg. The density of lead is 11 000 kg per cubic metre. Lead atoms are much heavier than water molecules and they are much more tightly packed together. So lead is eleven times denser than water (Fig. 1.5A).

RELATIVE DENSITY means the density of a substance compared to that of water. It can be calculated by dividing the density of the substance by the density of water. Relative density is also called SPECIFIC GRAVITY. See table 4, 'Relative density'. As an example, the RD or SG of aluminium is 2.7. This can be worked out in the following way: The density of aluminium is 2700 kg per cubic metre, and the density of water is 1000 kg per cubic metre. 2700 divided by 1000 = 2.7. Water is always shown as 1.0.

A substance will float on another substance if its density is less. Oil has an RD of 0.95 so it floats on water. Aluminium has an RD of 2.7 so will sink if put in water. The liquid metal mercury has an RD of 13.6. Lead has an RD of 11.3, which is less than the RD of mercury. Therefore, a lump of lead will float in a pool of mercury (Fig 1.5B).

A tonne of feathers weighs exactly the same amount as a tonne of lead. But the pile of feathers is very much bigger than the lump of lead. The lump of lead is more dense than the pile of feathers.

The specific gravity of liquids can be measured with a HYDROMETER (see 2.15).

The relative density of gases can be measured also. They are often compared with the RD of air which is shown as 1.0. Butane is more dense than air and therefore sinks to ground level. Helium is lighter than air and is used in balloons to make them float (Fig. 1.5C).

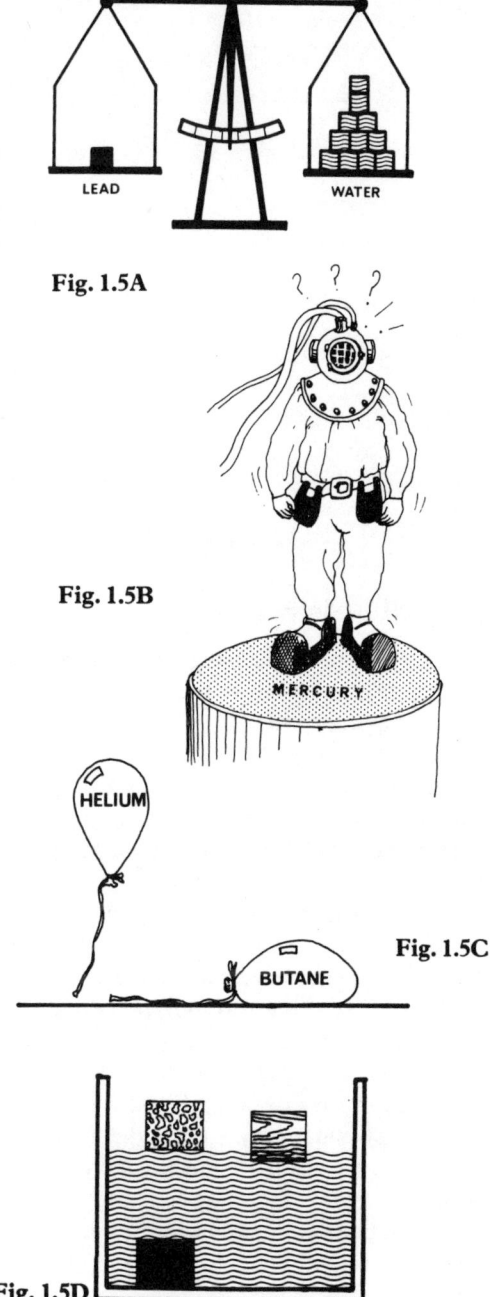

Fig. 1.5A

Fig. 1.5B

Fig. 1.5C

Fig. 1.5D

Table 4 Relative density (or specific gravity)
— *see Fig. 1.5D*

Substances	RD (SG)
Water	1.0
Cork	0.25
Deal	0.5
Alcohol	0.8
Linseed oil	0.95
Sand (dry)	1.6
Common brick	1.9
Concrete	2.4
Aluminium	2.7
Mild steel	7.7
Copper	8.9
Lead	11.3
Mercury	13.5
Gold	14.3

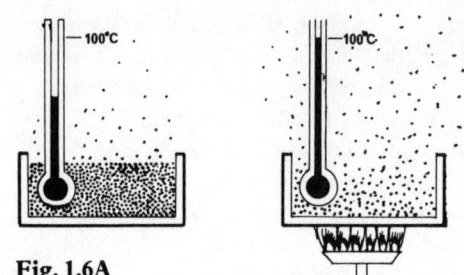

Fig. 1.6A

Movement of air across the surface of a liquid, such as wind across a puddle, speeds up evaporation.

Fluids like white spirit, resin thinners and petrol evaporate very easily at room temperatures. Because of this, spillages of FLAMMABLE liquids are dangerous as the room quickly fills with a mixture of flammable vapour and air. This mixture can be explosive (Fig. 1.6B)

1.6 What is evaporation?

EVAPORATION means changing a liquid into a VAPOUR, usually at below the temperature of the liquid's boiling point.

A puddle of water on a sunny day will gradually dry up. The water evaporates. The puddle of water does not boil away because its temperature does not reach 100°C.

Liquids evaporate because the MOLECULES on the surface have enough energy to break away. The surface molecules are not held in the liquid as strongly because they are only attracted by the molecules underneath. As the temperature increases the molecules get more energy. The molecules on the surface move more freely and break away more easily from the liquid. The hotter the day, the more quickly the puddle will dry.

Only at boiling point do all the molecules have enough energy to break free of each other (Fig. 1.6A).

Fig. 1.6B

1.7 What causes materials to expand and contract?

Depending upon the material there are two reasons: heat and moisture.

Most materials expand when heated. The heat causes the molecules to move about more vigorously. They move further apart and cause expansion.

When cooled, the molecules slow down, move closer together and the material gets smaller or contracts.

The mercury thermometer makes use of expansion and contraction. As the temperature rises the mercury expands and rises up the fine tube. As the temperature decreases the mercury contracts and drops in the tube. This expansion and contraction effect is very small. The tube is made with a fine bore so that the effect can easily be seen.

Materials that are POROUS and ABSORBENT can absorb moisture and expand as they take in the extra volume. As the liquid evaporates the material shrinks or contracts.

A dry piece of timber will expand if it absorbs a quantity of water. When it dries its pores and tubes contract.

There are many examples where these two conditions are met in the construction industry.

Bridges have to be constructed with expansion joints built in to allow for movements. This particularly applies to metal bridges.

Long runs of pipe in a hot water system must have expansion joints or bends in them to allow for expansion when hot water passes through them.

Flame cleaning uses a high temperature flame to expand millscale on steel. The thin layer of scale and the steel underneath expand at different rates when heated and adhesion between them is broken. This is known as DIFFERENTIAL EXPANSION.

Liquids expand a great deal when they are heated to boiling point, and form a gas. If bottles containing liquefied gases such as butane or propane or aerosols are heated, the liquid will expand into a gas and burst the container, causing a violent explosion.

The bottoms of doors and windows, if they are not sealed properly by painting, will absorb moisture, expand and make them difficult to open and shut.

When buildings are constructed a very large amount of water is used in the concrete, mortar and plaster. They may absorb more from rain and snow. Over a number of months this dries out and some shrinking or contraction will take place.

If damp or unseasoned timber is used in a building it will shrink as the moisture slowly evaporates. This can result in badly fitting doors and frames, also warping and panel shrinking.

When paint films dry, the solvents evaporate from the surface and the film shrinks or contracts. A dry paint film is always thinner than a wet film.

1.8 How do materials change their form?

Either by changing their physical form or by forming a completely new substance by a chemical reaction.

A substance may change its form but not its chemical properties. For example, water if frozen changes to solid water or ice. If boiled it changes to a gas known as steam. Its properties remain those of water even when a change in temperature changes its form.

When salt is dissolved in water a SOLUTION is formed. Neither the water nor the salt change their chemical properties and if the solution is boiled the water can be collected in the form of steam and the salt will be left behind in the bottom of the container.

A few paint films, such as shellac varnish, cellulose finishes and chlorinated rubber paints, can be softened by their own solvents and revert back to a liquid paint. These are REVERSIBLE COATINGS. Their physical form has been changed but their chemical properties remain unchanged.

Solder when cold is a solid. When heated it is a liquid. By raising its temperature solder can be changed from solid to liquid. When it cools it returns to a solid. This is a PHYSICAL CHANGE (Fig.

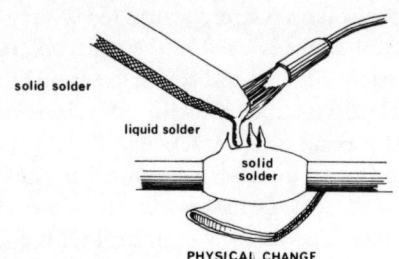

solid solder

liquid solder

solid solder

PHYSICAL CHANGE

Fig. 1.8A

1.8A). All physical changes can be reversed to obtain the original form of the material.

Other changes involve chemical reaction and an entirely new substance is formed with different chemical properties. There are very few CHEMICAL CHANGES which can be reversed.

When timber is burnt entirely new substances are formed. A gas is released in the form of smoke and fumes. Charcoal is left which has completely different properties to timber (Fig. 1.8B).

Iron corrodes and produces a weak scaly material known as rust.

Paints containing drying oils and resins dry by a chemical reaction into a solid film with entirely different properties to the original liquid paint.

None of these chemical reactions can be reversed to produce timber, iron or liquid paint.

Fig. 1.8B

wood is burnt and forms charcoal
CHEMICAL CHANGE

MOISTURE

2.1 What damage can moisture cause to buildings?

2.2 How does moisture move upwards through a wall?

2.3 What is condensation?

2.4 How can condensation be stopped?

2.5 What is efflorescence?

2.6 In what way do efflorescent salts and hygroscopic salts differ?

2.7 What happens to water when it freezes?

2.8 What happens to water when it is heated?

2.9 Is it necessary to know the moisture content of materials?

2.10 How can the moisture content of materials be measured?

2.11 Are all porous materials absorbent?

2.12 What is the most absorbent part of timber?

2.13 How does moisture affect the gloss and translucency of a surface coating?

2.14 What is hard water?

2.15 What is the difference between a hygrometer and a hydrometer?

SOURCES OF DAMPNESS

A. earth piled above d.p.c
B. blocked air vents
C. loose slates
D. blocked or broken r.w.p.
E. loose flashings
F. poor pointing
G. porous rendering
H defective cement plinth
J condensation

Fig. 2.1

2.1 What damage can moisture cause to buildings?

If unwanted moisture in a building is not stopped quickly it will reduce the life of the building and make it unhealthy to live and work in.

Dampness can damage buildings (Fig. 2.1) in many ways, including the following.

(a) Causing fungus growth which results in wet and dry rot in timber.
(b) Rusting of iron and steel.
(c) Reducing the adhesion of paints, plasters and adhesives.
(d) Staining caused by mould growth, condensation and colours in building materials.

(e) Condensation, mould growth, rotting timber and rising damp, resulting in smells and unhealthy atmospheres.
(f) Chemical actions on paints (saponification and bleaching) and on brickwork and plaster (efflorescence).

2.2 How does moisture move upwards through a wall?

Through the fine holes and gaps in many building materials.

A large number of common building materials contain fine holes and gaps. The holes in bricks and blocks are formed during manufacture. Holes form in concrete and plaster as they set.

Fig. 2.2A

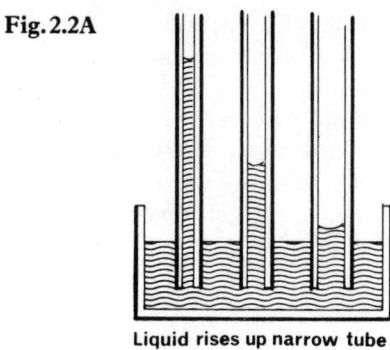

Liquid rises up narrow tube
by capillarity

Wood is full of natural holes which carried the moisture through the tree when it was growing. When these materials are in contact with the ground water enters these gaps. It is attracted to the material like metal to a magnet. The narrower the tubes and holes the higher the water rises. The movement of moisture is caused by CAPILLARITY, also called CAPILLARY ATTRACTION or CAPILLARY ACTION (Fig. 2.2A).

When building a wall this upward movement is stopped by putting a layer of material which does not have holes and tubes. This layer is called a damp-proof course and is put just above ground level but below floor level (Fig. 2.2B,). Damp

Fig. 2.2B

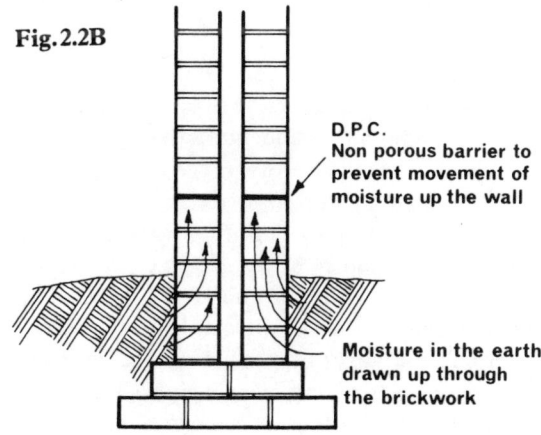

D.P.C.
Non porous barrier to prevent movement of moisture up the wall

Moisture in the earth drawn up through the brickwork

proof course materials may be slate, lead or bitumen felt. Without a damp-proof course walls may be affected by rising damp.

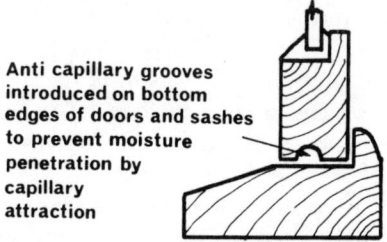

Anti capillary grooves introduced on bottom edges of doors and sashes to prevent moisture penetration by capillary attraction

Fig. 2.2C

2.3 What is condensation?

Water left on a cool surface by air which cannot hold any more moisture.

Air contains water vapour. The higher the air temperature the more water the air can hold. When air cannot absorb any more moisture it is SATURATED. If the temperature of air is lowered the water vapour will cool and some will turn to liquid water which forms on surrounding surfaces. This is called CONDENSATION (Fig. 2.3A).

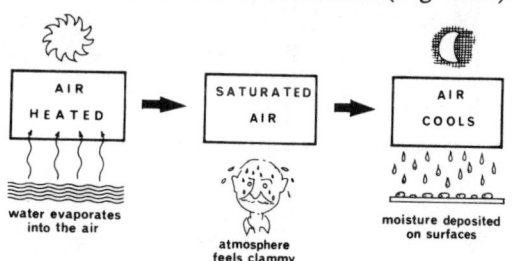

AIR HEATED → SATURATED AIR → AIR COOLS

water evaporates into the air

atmosphere feels clammy

moisture deposited on surfaces

Fig. 2.3A

The temperature at which water vapour in the air changes to liquid water is known as the DEW POINT.

Hot air does not necessarily contain water vapour. In desert areas where the temperature is high but there is very little water the air will be hot and dry. Whereas in a shower room the air temperature is high also but contains a large amount of water vapour. This makes the air feel clammy. The water vapour content of air is called HUMIDITY. A shower room would be of high humidity.

The amount of water vapour in the air can be measured. This measurement is called RELATIVE HUMIDITY. It is measured with an instrument called a HYGROMETER.

After a hot sunny summer's day the air cools. Condensed moisture forms on surrounding surfaces – this is called dew.

SOURCES OF MOISTURE IN HOMES

Fig. 2.3B

In houses moisture in the air is caused by cooking, bathing and breathing (Fig. 2.3B). Each person breathes out the vapour from one litre of water every day. More condensation is likely to form in winter because doors and windows are kept closed to contain the warmed air. Warm, still air holds a lot of moisture. When this saturated air comes in contact with a window or glazed tiled wall, which is chilled by its contact with the outside cold air, the air is cooled to its dew point. The water will be deposited on that surface in the form of condensation.

Condensation forms very quickly at night when the house heating is switched off. The air which held a lot of moisture when warm quickly cools to its dew point and gets rid of its moisture on any cold, hard surface.

2.4 How can condensation be stopped?

Change the air by improving ventilation, and raise the temperature of the interior surfaces by insulation.

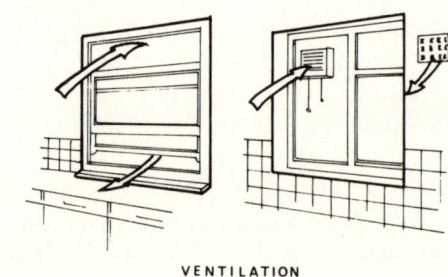

VENTILATION

Fig. 2.4A

Ventilation (Fig. 2.4A)

There are three ways of ventilating a building to reduce condensation. One is to open windows to replace damp air inside with drier air from outside. The second is to use extractor fans which change the air by drawing out the damp air and replacing it with dry air. Both of these methods also reduce the air temperature by changing the warm air inside with cooler air from outside. The third method is to use air conditioners, sometimes called DEHUMIDIFIERS. These extract moisture from the air and pass the air back into the room without lowering its temperature.

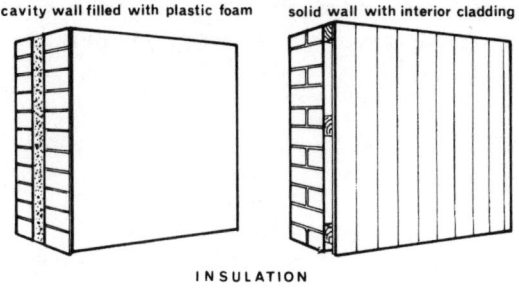

cavity wall filled with plastic foam   solid wall with interior cladding

INSULATION

Fig. 2.4B

Insulation (Fig. 2.4B)

If the temperature of surfaces is raised condensation is less likely to form. The best material for insulation is a poor conductor of heat, such as still air. Therefore processes are used which trap still air between the cold surface and the atmosphere. Materials like cork and foams are good insulators because they contain many pockets of still air. Insulators are used in the following ways.

(a) Filling wall cavities with plastic foam or fibres which prevent heat from the room passing through the inner wall.
(b) Lining walls with expanded polystyrene.
(c) Cladding walls with wood panelling or cork tiles.
(d) Using paints containing particles of cork, exfoliated vermiculite or perlite.

Condensation on windows can be reduced by double glazing. Very little heat travels between the two sheets of glass as the air trapped inside acts as an insulator.

2.5   What is efflorescence?

EFFLORESCENCE is a fluffy white powder which appears on the surface of new brick or plaster walls.

The powder is made when CRYSTALS which contain water dry out on the surface. The crystals are chemicals which may be in the bricks themselves, in the sand used in the mortar, or dissolved in water which has crept up the walls from the ground.

When mortar or plaster is mixed the water dissolves the chemicals. When they dry out the chemicals are drawn towards the surface. The water in these crystalline chemicals evaporates into the air. The remainder is an unsightly white powder on the surface of the walls.

Efflorescence forms on new external brickwork and cement renderings. The action stops when the walls are dry. The salts may not damage the wall but look unsightly. In time the weather removes them.

If internal walls are decorated before efflorescence has stopped paints and wallpapers will be pushed off as the crystals form. Any efflorescence left on a surface before decorating will stop adhesion and cause peeling or flaking.

Old walls sometimes show efflorescence because of dampness entering them. This shows there is a fault with the structure that must be put right.

2.6   In what way do efflorescent salts and hygroscopic salts differ?

EFFLORESCENT salts give off water. HYGROSCOPIC salts absorb water.

Efflorescent salts

Efflorescent salts lose water leaving behind a powder. These salts are CRYSTALS. The crystals are made of two parts joined together, a chemical and water. The water is called WATER OF CRYSTALLISATION. When these crystals are left in the open air they break down. The water in them evaporates and the crystals turn to a powder.

If washing soda crystals are left in an open packet water is slowly lost from the soda and a white powder remains (Fig. 2.6A).

Efflorescent salts form on the surface of new brick, concrete or cement rendered walls.

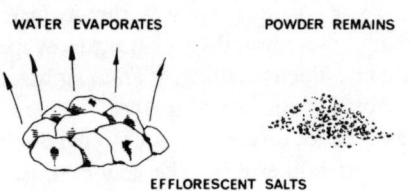

WATER EVAPORATES   POWDER REMAINS

EFFLORESCENT SALTS

Fig. 2.6A

Hygroscopic salts
Air contains water in the form of water vapour, hygroscopic salts attract this water from the air. Common salt is hygroscopic. The salt gets damp if its container is left open. Some solids can absorb enough moisture to form a solution. These substances are DELIQUESCENT.

If a tray containing dry pellets of caustic soda is left in the open air, after a few days there will be a pool of water in the tray in which the caustic soda has dissolved (Fig. 2.6B). Caustic soda is deliquescent.

If materials used in building, like plaster, contain hygroscopic salts they will attract water and stay damp. This happens in buildings which have been flooded with sea water. When the

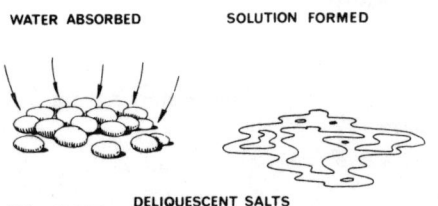

WATER ABSORBED   SOLUTION FORMED

DELIQUESCENT SALTS
Fig. 2.6 B

water dries out the salts left are hygroscopic and the building continues to absorb moisture from the atmosphere.

2.7 What happens to water when it freezes?

When water freezes it becomes ice.

When ice is formed the water turns from a liquid to a solid. This is known as a PHYSICAL CHANGE because when ice is melted it changes back to liquid water. Water turns to ice at a temperature of 0°C. Ice is made of crystals.

Ice is not as dense as liquid water, that is why it floats on the surface of ponds (Fig. 2.7). When water freezes it expands slightly and takes up more room than it did as liquid water.

Fig. 2.7

The expansion of ice can do a lot of damage. In winter water pipes may freeze, opening joints in pipes or splitting them. Car engine blocks may crack when the cooling system freezes.

If water gets into bricks and freezes, the brick can crack or *spall*, forcing off the face of the brick.

2.8 What happens to water when it is heated?

When water is heated it expands by a small amount. Central heating systems which use water in radiators have an expansion tank into which water can go to prevent bursting.

When water is boiled it turns from a liquid to a gas called steam. This is another example of a PHYSICAL CHANGE because the steam turns back to water when it cools. Water boils at a temperature of 100°C at sea level. There is an enormous expansion when water turns to a gas. The steam from a kettle of water can fill several rooms (Fig. 2.8).

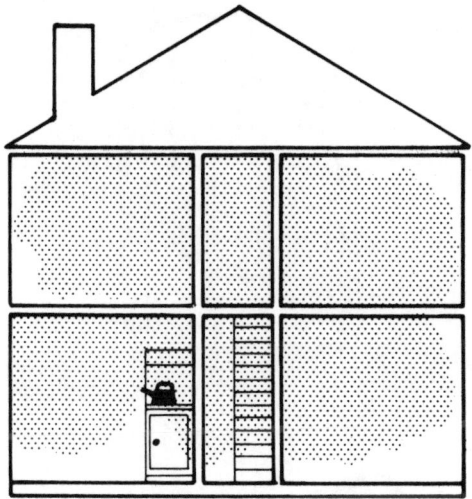

Fig. 2.8

2.9 Is it necessary to know the moisture content of materials?

Yes, because water in materials can prevent adhesion, or cause rotting and swelling.

The amount of water contained in a material can affect its performance when used in a building. The size of wood is increased when it contains water and if this water evaporates the wood shrinks, splitting and distorting its shape.

Wood containing large amounts of water may rot.

The adhesion of paints to a material can be weakened by water on or near its surface preventing the paint from forming a strong mechanical key.

Adhesives used for sticking tiles to walls or floors will lose adhesion or be weakened by water in the surface.

2.10 How can the moisture content of materials be measured?

By using either a simple piece of equipment called a MOISTURE METER, or by cutting and weighing part of the material.

The meter is most commonly used. It is a simple, easily carried piece of equipment which can be used to measure the moisture content of plaster, concrete or timber on site (Fig. 2.10A).

The meter works by passing a low voltage electrical current through the surface. Because impure water is a good CONDUCTOR of electricity more current passes through a wet surface than a dry one. The amount of current is shown on a dial or on a lit display. The numbers usually show the percentage of water in the material. For example, a 10% MOISTURE CONTENT means that for every 100 cubic millimetres of material there is also 10 cubic millimetres of moisture.

The most accurate type of meter has two metal spikes which are pushed into the surface and the electrical current passes between their points.

Another type has a pad which is pressed on the surface and the current passed along it. This measures surface wetness only, which may be very different to the moisture below the surface.

Fig. 2.10A

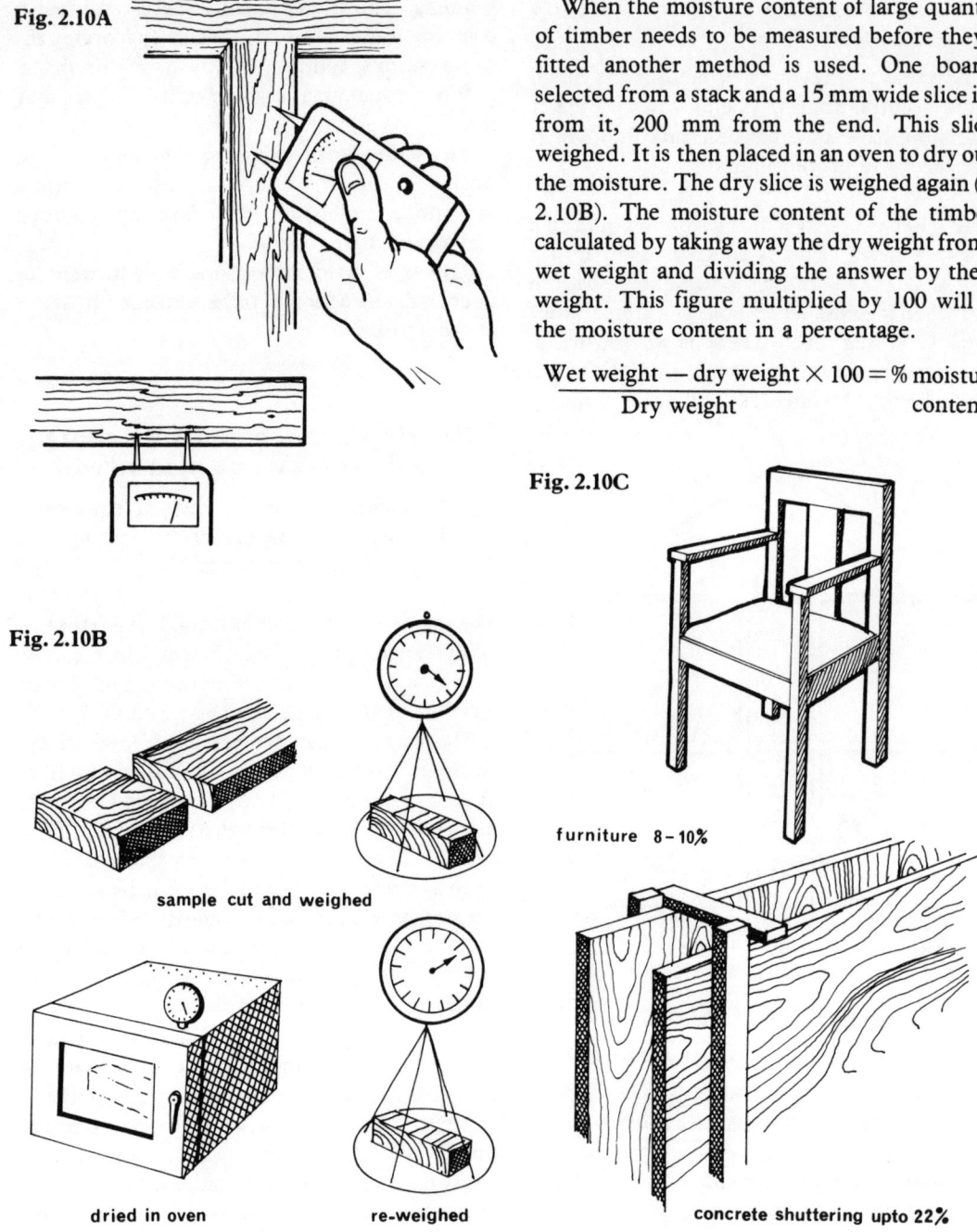

When the moisture content of large quantities of timber needs to be measured before they are fitted another method is used. One board is selected from a stack and a 15 mm wide slice is cut from it, 200 mm from the end. This slice is weighed. It is then placed in an oven to dry out all the moisture. The dry slice is weighed again (Fig. 2.10B). The moisture content of the timber is calculated by taking away the dry weight from the wet weight and dividing the answer by the dry weight. This figure multiplied by 100 will give the moisture content in a percentage.

$$\frac{\text{Wet weight} - \text{dry weight}}{\text{Dry weight}} \times 100 = \% \text{ moisture content.}$$

Fig. 2.10C

furniture 8 – 10%

Fig. 2.10B

sample cut and weighed

dried in oven

re-weighed

concrete shuttering upto 22%

Timbers are required to have different moisture contents depending upon the use to which they are put.

Wood for furniture needs a moisture content of about 8 to 10% (Fig. 2.10C). A centrally heated house needs timber with a moisture content of about 12 to 14%. In offices where central heating is much hotter and is left on longer than in homes, the moisture content may need to be as low as 8%. For roofs or wall framing, or for concrete shuttering, timber with a moisture content of up to 22% can be used. When the moisture content is more than 20% the wood rot fungus is most likely to start feeding on the wood.

2.11 Are all porous materials absorbent?

Only if all the pores can be filled with water.

A material is POROUS if it contains very small holes, called PORES. Bricks, plaster, wood, cork, sponge and foamed plastic are porous. The pores are full of air.

If the air in the pores can be replaced with water the material is ABSORBENT.

In some materials these pores join together. When water is absorbed it passes from pore to pore, usually by CAPILLARITY, pushing the air out as it goes. These materials are absorbent and include brick, plaster and timber (Fig. 2.11A). If

the water can pass right through a material it is called PERMEABLE.

Foamed polyurethane slabs and cork are full of pores, but they are not connected so cannot pass the water on. These materials are porous but NON ABSORBENT. For this reason they float and are used to give extra buoyancy in boats, and for safety jackets (Fig. 2.11B).

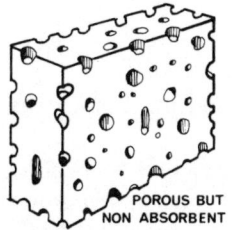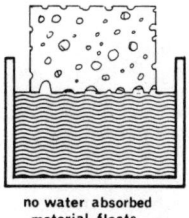

POROUS BUT NON ABSORBENT

no water absorbed material floats

Fig. 2.11B

In bricks and foams the pores are formed when they are made. In wood and cork they occur naturally.

Some materials are completely solid and have no pores. Water cannot be soaked up by them because it has nowhere to go. These materials are both non porous and non absorbent and include the metals and glass.

2.12 What is the most absorbent part of timber?

The end grain because it is a mass of very open pores.

When trees are growing moisture passes up through the roots in fine pipelines to the branches and leaves. Without this moisture the tree would die. When the tree is cut down and dried the moisture in these pipelines EVAPORATES turning them into fine CAPILLARY tubes (Fig. 2.12A). When trees are cut into planks these tubes are exposed as fine holes at the end of each plank,

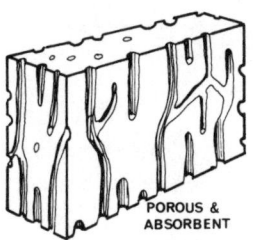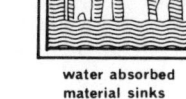

POROUS & ABSORBENT

water absorbed material sinks

Fig. 2.11A

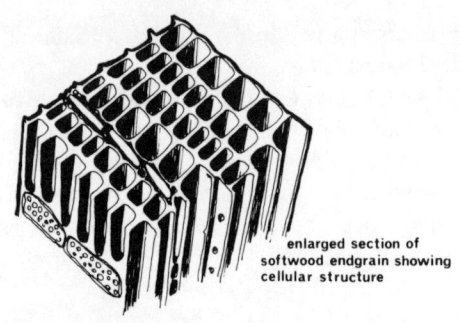

enlarged section of
softwood endgrain showing
cellular structure

Fig. 2.12A

making moisture penetration much easier than on the face or edges.

When used in the construction of a building it is important that this end grain is well protected from moisture. Special attention should be paid to these areas when priming window and door frames before they are fitted into brickwork (Fig. 2.12B).

The ends of timber floor joists should be wrapped in waterproof building paper before fixing to stop the wood absorbing moisture from the brickwork, which may cause rot.

Where end grain is not painted, like ends of

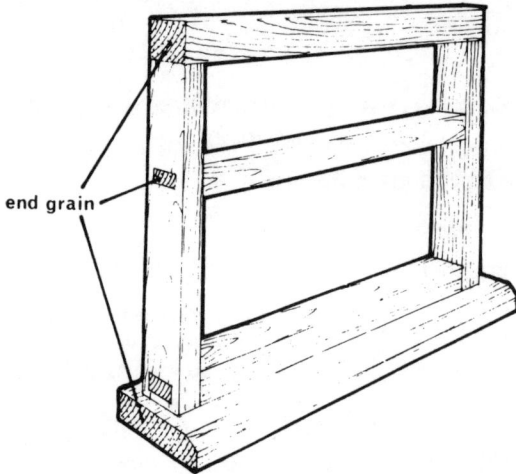

end grain

Fig. 2.12B

timber joists and rafters it should be treated with timber preservative. This will prevent wood rot if the timber absorbs a lot of water.

2.13 How does moisture affect the gloss and translucency of a surface coating?

It can make gloss paints appear dull and varnishes cloudy.

If moisture settles on a paint film before it has dried it will sink into the film making minute holes (Fig. 2.13A). When the paint dries the surface will be left with a mottled texture. This texture affects the amount of gloss the surface will have.

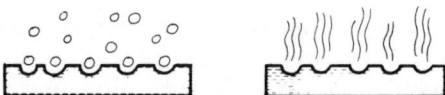

Fig. 2.13A

The amount of gloss a surface has depends on the amount of light we see reflected from it. When light strikes a smooth surface it is reflected from the surface as from a mirror (Fig. 2.13B).

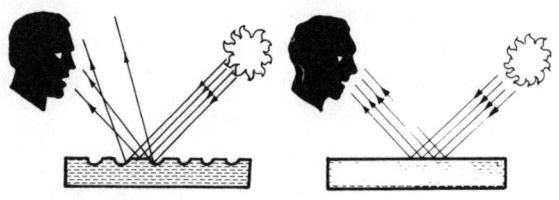

Fig. 2.13B

When the surface is not smooth the light will be reflected at many angles, giving the appearance that the gloss is less shiny because less light will be seen by the eye. An effect like this may be seen by bouncing a ball. When it is bounced on a

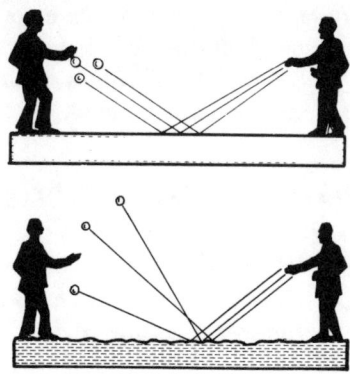

Fig. 2.13C

smooth surface the bounce is true. When it is bounced on an uneven surface it may go in any direction (Fig. 2.13C).

Another form of dullness can appear on dry gloss paint and is called blooming. The appearance of blooming is similar to the dull film which occurs on the skin of some fruits, particularly grapes. This is caused when moisture containing impurities from the air CONDENSES on the drying paint film. When the water evaporates it leaves behind tiny crystals of the impurities on the surface which give a dull cloudy look. Sometimes the crystals causing the blooming can be wiped off when the paint is dry. In some cases the impurities may eat into the surface making deep pits and the gloss cannot be recovered.

When blooming happens on varnishes they lose their TRANSLUCENCY and look like thin white paints.

2.14 What is hard water?

Water is called hard when it will not form a lather with soap and the use of more soap only forms a scum. The hardness is caused by various chemicals which have been dissolved in the water.

When rain falls through the air it dissolves carbon dioxide and forms a weak acid. When it reaches the ground this weak acid dissolves various solids as it passes over and through the soil on its way to the reservoirs. The amount and kinds of impurities dissolved depend on the type of local soil.

There are two kinds of hard water, called *temporary hardness* and *permanent hardness*.

Temporary hardness is generally caused by calcium bicarbonate in the water. This happens when the water containing the carbon dioxide passes over chalk or limestone.

The hardness can be removed by boiling the water. Boiling releases the carbon dioxide and deposits the chalk again. The water which is left will form a good lather with soap. Unfortunately, the chalk which drops out forms hard coatings on the inside of kettles, boilers and hot water pipes, and slowly chokes them up. This deposit is called fur or scale (Fig. 2.14A).

Reservoirs can remove most of the *temporary hardness* by adding slaked lime to the water. The method by which this is done is called the *Clarks process*.

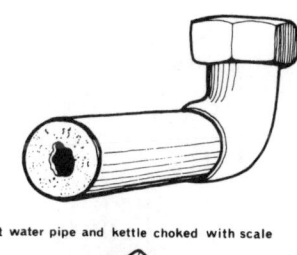

hot water pipe and kettle choked with scale

Fig. 2.14A

Permanent hardness is caused by calcium or magnesium sulphate dissolved in the water. It cannot be removed by boiling and for this reason it is called permanent.

It can be cured by adding washing soda to the water. The soda joins with the sulphates and chalk is formed and drops out of the water. This is known as PRECIPITATION.

Where large quantities of water need softening, such as in laundries, special processes are used. These are called permutit or zeolite processes.

Permutit is a container of special chemicals. When the water is run through it, the chemicals decompose the calcium salts, leaving the water soft (Fig. 2.14B). At regular intervals the system is flushed through with common salt to renew the chemicals.

Water which makes lather easily is called soft water. It is usually found in areas where the ground is mainly rock and contains no chalk.

Hardness can be measured by the amount of lime in the water. Eight degrees of hardness is considered to be soft. Water with a reading of above eight is considered to be hard.

2.15 What is the difference between a hygrometer and a hydrometer?

A hygrometer measures the RELATIVE HUMIDITY of air. A hydrometer measures the RELATIVE DENSITY of a liquid.

There are many types of hygrometers (Fig. 2.15A). One type is like a pocket watch which contains many holes to let the air get in. A coil spring of paper inside changes its shape depending on the amount of moisture in the air. As it moves, the spring works a pointer which indicates the RH as a percentage.

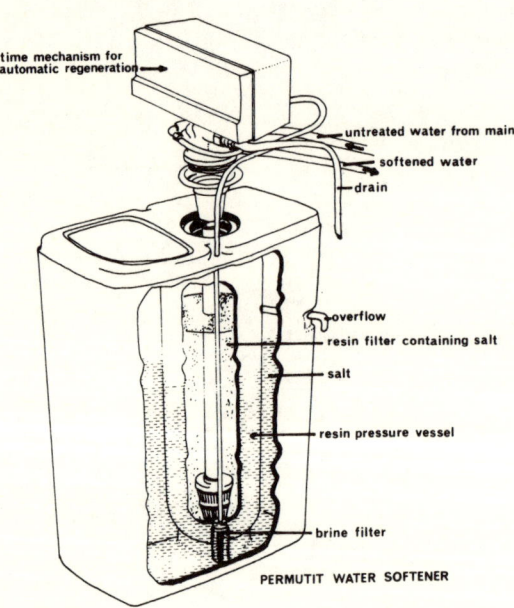

Fig. 2.14B

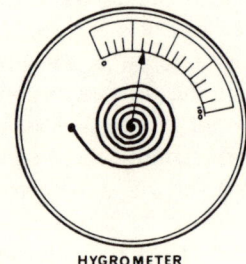

HYGROMETER

Fig. 2.15A

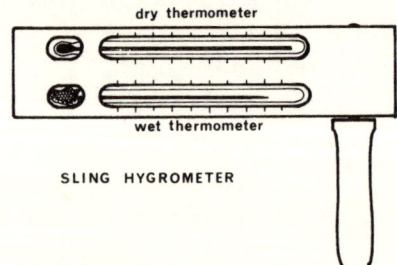

SLING HYGROMETER

Fig. 2.15B

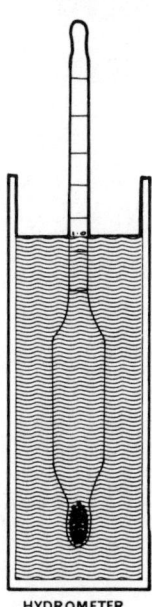

HYDROMETER

Fig. 2.15C

A sling or whirling hygrometer is sometimes used (Fig. 2.15B). This is whirled round at speed. It is called a ventilated bulb instrument. Readings have to be taken from the instrument after it has been whirled and the RH can be calculated from the two figures by use of a graph.

The most common type of hydrometer is like a float (Fig. 2.15C). It is a glass tube with a weighted end. When floating in water the mark 1.0 on the tube will be at the level of the water. In other liquids it will either sink deeper and record a lower number than 1.0, or float higher and record a figure more than 1.0. In alcohol it will sink deeper because alcohol has a lower RD than water.

ADHESION

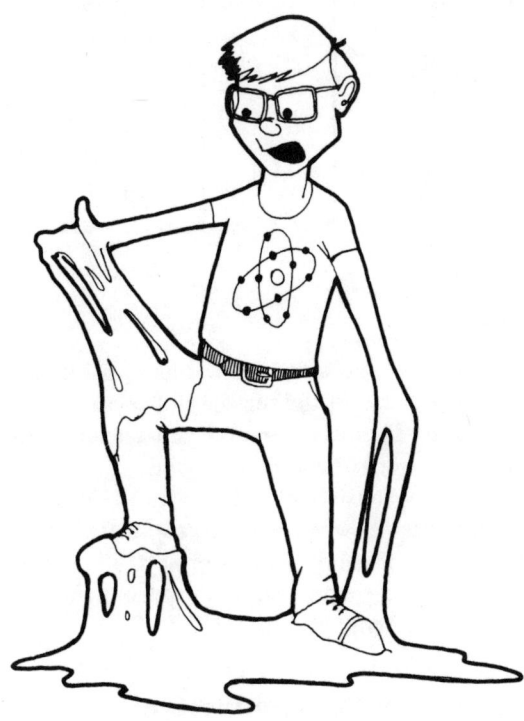

3.1 What are adhesives and how do they work?

3.2 How can the adhesive properties of a surface be improved?

3.3 Why does an adhesive sometimes break down?

3.4 How does plaster adhere to plasterboard?

3.5 How does paint adhere?

3.6 What is soldering?

3.7 Why may paint flake from a dry absorbent surface?

3.8 Why may a wet surface cause paint to blister and peel?

3.9 How is grease removed from a surface?

3.1 What are adhesives and how do they work?

Adhesives are substances which will stick different materials to each other so that they will remain joined indefinitely.

Very many adhesives are available to suit the many different materials that require joining. Adhesives vary from simple starch pastes for sticking wallpaper to plaster walls, to complex chemical combinations needed to bond components on space vehicles.

Adhesives contain various ingredients, the most important being the binder which makes the adhesive stick or ADHERE to the surface, and holds the particles of the adhesive together. These are its ADHESIVE and COHESIVE properties. Other additives may be used also. These include: HARDENERS, which chemically CURE or dry the adhesive; pigments to colour the adhesive so that it matches the materials; PLASTICISERS to make the adhesive more flexible; THIXOTROPIC agents to thicken the adhesive so that it will not sag on vertical surfaces; and wetting agents to make sure that the adhesive has very close contact with the surface.

Many substances can hold materials together but can only be considered as adhesives if they make a permanent bond. If water is frozen between two pieces of glass they will be joined only as long as the water stays frozen so the water would not be termed an adhesive.

The mechanism of the bonding between the adhesive and the surface to be bonded is complicated. A simple way of explaining it is that it is a combination of SPECIFIC and MECHANICAL ADHESION (Fig. 3.1). Specific adhesion is the attraction between the molecules of the adhesive and the molecules of the surface, plus the cohesive strength of the molecules in the adhesive. Mechanical adhesion is the anchoring of the adhesive in the texture of the surface. In

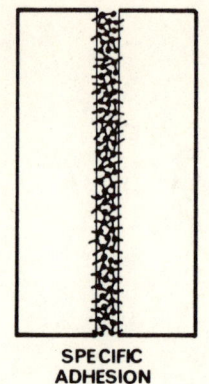 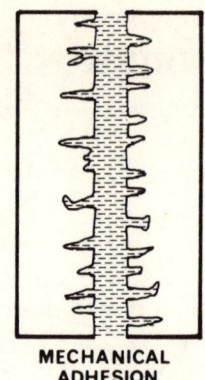

SPECIFIC ADHESION   MECHANICAL ADHESION

Fig. 3.1

other words, the adhesive hardens inside the pores and cracks of the surface and keys itself in. The greater the surface areas in contact with each other the stronger will be the bond.

Adhesion consists of three stages. First the liquid adhesive, when applied, completely wets the surface. Second, it dries or sets. Finally, it forms a strong permanent bond without altering the surfaces being bonded.

3.2 How can the adhesive properties of a surface be improved?

By making sure that as much as possible of both surfaces are touching and that there is no grease, dirt or water between them.

When two different materials stick together it may be the result of MECHANICAL ADHESION, SPECIFIC ADHESION or a combination of both. The strongest bonds are obtained when the greatest areas of both surfaces are in contact. Very rough surfaces will have few points of contact and must be broken down to increase surface contact (Fig. 3.2A).

Moisture can form a barrier between the adhesive and the surface, preventing or reducing

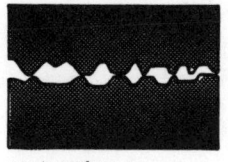

rough surface

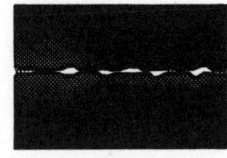

smooth surface

Fig. 3.2A

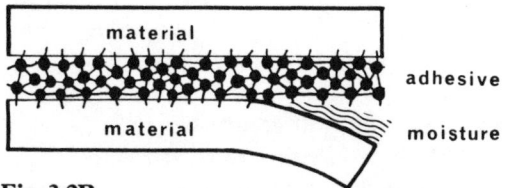

material

adhesive

material

moisture

Fig. 3.2B

contact, and must be removed (Fig. 3.2B). Grease, oil and dirt on the surface will lessen the wetting properties of adhesives with the surface. Very smooth shiny surfaces should be abraded to provide a mechanical key for paint coatings or adhesives. This is important when applying wallpaper to gloss painted walls. Before applying paint to smooth surfaces such as aluminium an etch primer should be applied. This will roughen the surface slightly and provide a mechanical key.

3.3 Why does an adhesive sometimes break down?

The most common cause is the use of the incorrect adhesive for the work in hand (Fig. 3.3).

Moisture, strong solvents, heat or mechanical stress may cause an adhesive to break down.

The heating of copper pipes close to a soldered joint may result in the solder melting and breaking the bond. Moisture contained in walls covered with wallpaper may cause softening of the paste. Adhesives which contain organic ingredients such as casein or glue may decompose

if attacked by mould growths. Incorrect or poor preparation of the surfaces to be bonded will result in the adhesive not wetting the surface completely and will prevent close contact between the surface and the adhesive. This is very important with the thin *instant* adhesives. Two-pack adhesives will not cure or harden if mixed in wrong proportions or if the surrounding temperature is too low, or the air is wet. Glue is brittle and will break into small pieces if moved a lot or struck a heavy blow.

Fig. 3.3

3.4 How does plaster adhere to plasterboard?

When plaster is mixed with water, crystals are formed which interlock with themselves and the surface of the plasterboard forming a strong bond.

Crystalline gypsum, calcium sulphate, is heated

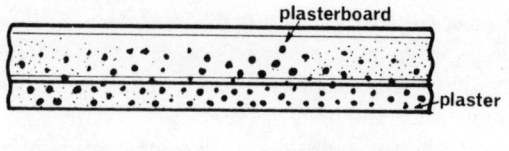

Fig. 3.4

during manufacture; this removes its WATER OF CRYSTALLISATION leaving a dry powder. This powder is called anhydrous gypsum plaster.

When water is added to plaster the calcium sulphate dissolves making a SOLUTION. As plaster sets CRYSTALS are formed which interlock to form a strong crystalline structure. When wet plaster is applied to plasterboard some of the saturated solution is absorbed by the board. As the solution sets crystals are formed. The crystals interlock with the plaster in the board and also with the crystals being formed on the surface of the board (Fig. 3.4).

It is important that the surface of the board is clean for good absorption of the plaster into the board and for this reason it is important that the plaster is applied as soon as possible after the board has been fixed.

Many plasterboards have a different paper on each face. One of them allows the wet plaster to be absorbed fully to obtain a good bond. The other is sealed for easy painting but poor plaster bonding. It is essential that the right face is used for good adhesion.

3.5 How does paint adhere?

By absorption, stickiness or by softening the previous coatings.

When paint is applied to unpainted porous materials such as wood or plaster some of the paint will be absorbed into the surface. When dry the absorbed paint helps to anchor the coating to the surface. On non-absorbent surfaces, such as glass, this anchoring is not possible and the strength of adhesion will rely on the *stickiness* of its binder to hold the film on to the surface. This is SPECIFIC ADHESION. It is for this reason that on smooth surfaces, like glass which cannot be given a mechanical key, gloss paint or varnish is used as a first coat. Gloss paint has a greater proportion of binder than undercoat and so there will be more binder in contact with the surface and therefore a stronger adhesion will occur.

The adhesion of paint over existing coatings may be improved by the softening action of its solvent during and just after application. If the solvents in the freshly applied paint are strong enough to soften the binder in the underlying coating there will be a degree of fusion between them. This will only occur before the solvent evaporates. If this softening takes place the paints are termed REVERSIBLE. Cellulose paints and chlorinated rubber paints are reversible and a good bond is obtained between coats.

Softening of coatings can occur with air drying oil paints to a certain extent if the paint is applied before the OXIDATION of its binder is complete. The increase of adhesion by solvent action is important when applying second coats of some two-pack paints or varnishes. Due to the hardness of the cured coatings it is important to apply the second coat within a specified time, or the surface will need to be etched to make a mechanical key for the adhesion of another coat.

3.6 What is soldering?

Soldering is a way of joining pieces of metal together by means of another metal, solder, which can be melted very easily.

Solder has a low melting point. In its molten state

it will flow on and into the parts of the metals to be joined. When it cools and turns back into a solid it forms a strong bond between the metals.

Solder conducts electricity and is used largely in the electrical industry for joining wires together. Most common solders are alloys of tin and lead. These alloys usually have a melting point under 430°C.

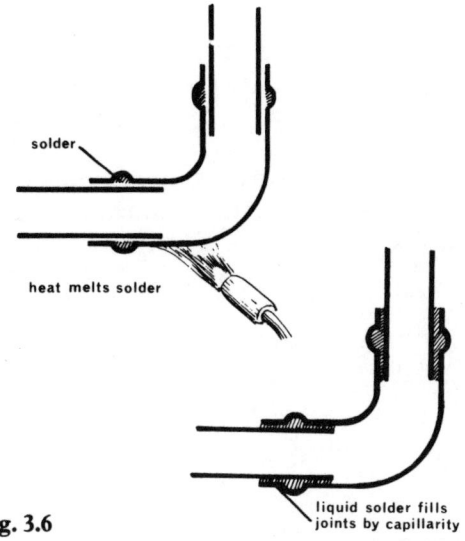

Fig. 3.6

solder

heat melts solder

liquid solder fills joints by capillarity

Solder is commonly used for the joining of copper pipes in domestic hot water and cold water systems as the temperatures of the pipes do not reach the melting point of solders (Fig. 3.6).

To help adhesion the surface of the metals to be joined must be clean. When heated the solder quickly melts and flows into the joints by capillarity. It then spreads and attaches itself to the surfaces of the metals. Chemicals such as flux are usually applied to the surfaces to stop the formation of contamination which would prevent a strong bond being made. A weak bond is known as a dry joint. Electrical solder often has flux already inside it, in a core.

3.7 Why may paint flake from a dry absorbent surface?

Because the liquid part of the paint is absorbed leaving a poorly bound film.

When paint is applied to a dry absorbent surface, much of the thinner and binder will be drawn into the surface. As the thinner is quickly absorbed the coating will lose its wetness and cannot be spread evenly. This will result in areas of thick underbound paint giving an uneven coating on the surface. The coating has only a weak bond with the surface and will lack FLEXIBILITY because it has lost a lot of its elastic binder.

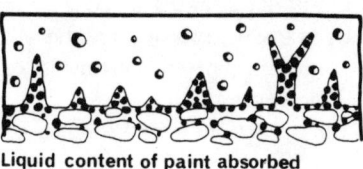

Liquid content of paint absorbed

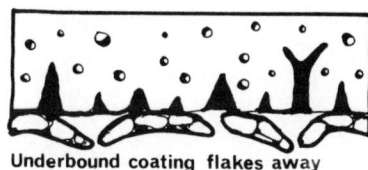

Underbound coating flakes away

Fig. 3.7

Surface movement or the weight of extra paint coatings will put stresses on the dried film. This may cause the film to crack if it is not flexible enough to withstand these stresses (Fig. 3.7). After cracking the film may flake off if its adhesion is not good enough to hold it to the surface. For these reasons sealers are made which contain a sufficiently high proportion of binder and thinner to satisfy the POROSITY of absorbent surfaces.

3.8 Why may a wet surface cause paint to blister and peel?

Because water prevents paint adhering to a surface.

If a dried paint film is to remain firm on the surface to which it is applied, it must adhere to the *whole* of that surface and, if possible, penetrate it. Moisture on or in the surface will form a barrier preventing adhesion taking place. When paint is absorbed into the pores of a surface by CAPILLARITY it mechanically anchors into the surface forming a strong bond. If the surface contains moisture this will not happen. Moisture on the surface will further reduce the bond as the paint will not have intimate contact with the surface. Its SPECIFIC ADHESION will be weak. As it dries the paint film contracts further so weakening the adhesion even more.

When the dried film later becomes hot it will expand (Fig. 3.8). If its adhesion is poor in places where moisture has prevented a good bond the film will expand into dome shaped blisters. In bad cases of poor adhesion the whole film may peel away from the surface.

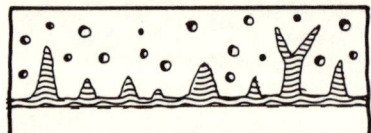

Moisture trapped in surface

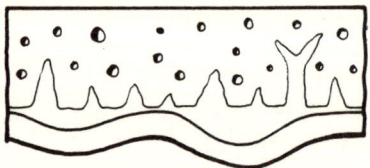

Heat pushes up paint film

Fig. 3.8

3.9 How can grease be removed from a surface?

By use of solvents, detergents and emulsifying agents.

The particles which make up grease or fat are arranged close together in a semi-solid state. This is called a COLLOID. The grease can be dissolved by introducing a suitable solvent which will cause the tightly packed particles to move around more freely in a liquid state. In this liquid state the grease can be wiped from the surface with absorbent paper or rags.

Hydrocarbon solvents, such as white spirit, are good degreasers but they give off flammable vapours as they evaporate. Chlorinated hydrocarbons, such as trichlorethylene, are non-flammable and are widely used as degreasing

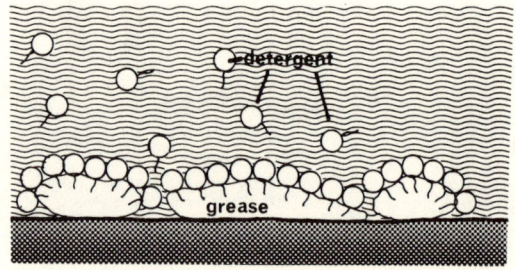

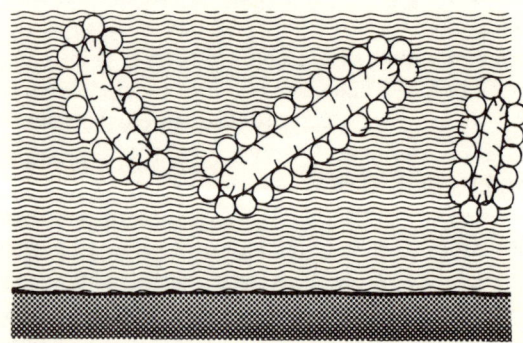

Fig. 3.9

agents. Their main disadvantage is the toxic vapour they give off, which is worse on heated surfaces.

The vapour of some solvents can be used to degrease. Vapours condense on the greasy surface, dissolve the grease and run off.

Detergents consist of tiny particles shaped like tadpoles. These particles are called SURFACTANTS. Their head is attracted to water and their tail is attracted to grease and dirt. When dispersed in water the surfactants move around rapidly attaching themselves to grease and dirt (Fig. 3.9). Eventually the surfactants form long tubes with the dirt and grease trapped inside. Rinsing with clean water removes the detergent along with the dirt and grease, leaving the surface clean.

Emulsifying agents are non-detergent soaps, such as sugar soap, which are made by reacting fatty acids with an alkali to form a neutral soap. When dissolved in water and applied to a greasy surface the soap combines with the grease to form an EMULSION which can be washed off the surface with clean water, leaving the surface free from grease.

HEAT

4.1 What is meant by temperature?

4.2 How is temperature measured?

4.3 Why do some things feel hotter when they are being warmed by the same source of heat?

4.4 What is meant by the thermal conductivity of a material?

4.5 Is the air hotter above a radiator than below it?

4.6 How does a hot water radiator system work?

4.7 How does the heat get from the sun to us?

4.8 What type of materials heat up slowly?

4.9 Is heat reflected?

4.10 How can heat be contained in a building?

4.11 Does a dark coloured surface become hotter than a light coloured one?

4.12 Do all materials heat up and cool down at the same rate?

4.13 Is one fuel as good as another for producing heat?

4.14 What causes walls and ceilings above a radiator to get dirty?

4.15 What is pattern staining?

4.16 What is refrigeration?

4.17 What is a thermostat?

4.18 What is latent heat?

4.19 Is black a good colour to use on a roof?

4.1   What is meant by temperature?

It is a way of describing how hot or cold a substance is.

To describe more clearly what hot and cold mean temperature scales are used. The most common scale is CELSIUS (centigrade). This uses zero degrees (0°C) as the TEMPERATURE at which water freezes, and 100°C as the temperature at which water boils. Temperatures below freezing are shown as minus degrees Celsius (Fig. 4.1). Temperatures above boiling point will be more than 100°C. By using a scale of this sort one temperature can be compared with another.

Another temperature scale which is used by scientists is called the KELVIN scale. It starts at 0°K which is ABSOLUTE ZERO and is equivalent to −273°C. Boiling point of water on the Kelvin scale is 373°K.

4.2   How is temperature measured?

Temperature is measured with a thermometer.

Common THERMOMETERS are sealed thin glass tubes with a bulb at one end. They contain either alcohol, which is usually coloured red, or mercury, which is silver. When alcohol or mercury are heated they expand and travel up the tube. The points at which they rest when the thermometer is in contact with ice is marked 0, and with boiling water is 100. The space in between is divided into 100 equal parts and each one is called a degree. This temperature scale is called CELSIUS.

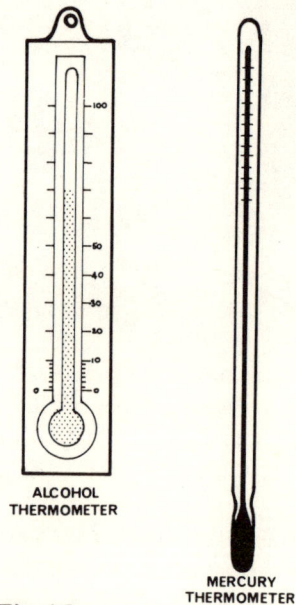

Fig. 4.2

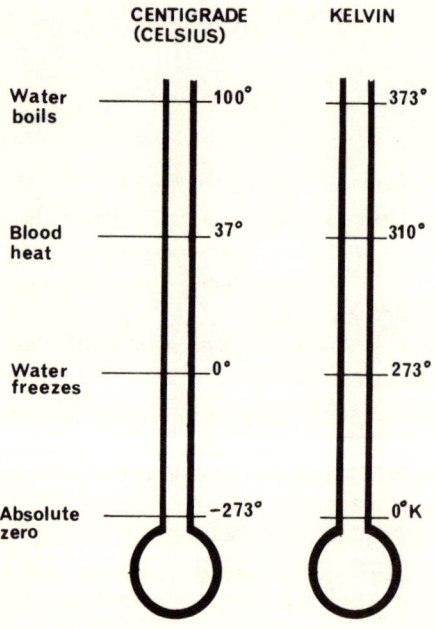

Fig. 4.1

Alcohol thermometers (Fig. 4.2) are commonly used for measuring room temperatures. They are cheaper than mercury thermometers. The alcohol also expands more and travels up the glass tube further so that the thermometer is easier to read.

Mercury thermometers are commonly used in

laboratories where temperatures to be measured may be above 100°C because alcohol boils at 80°C. But alcohol thermometers are used for measuring very low temperatures, because mercury freezes at —40°C.

4.3 Why do some materials feel hotter when they are being warmed by the same source of heat?

Because some materials allow heat to pass through them quickly, while heat passes through others very slowly.

Heat passes through materials at different speeds. Most metals allow heat to pass through them quite quickly. Most non-metals allow heat to pass through them slowly. Materials which pass heat slowly are called poor conductors of heat or INSULATORS.

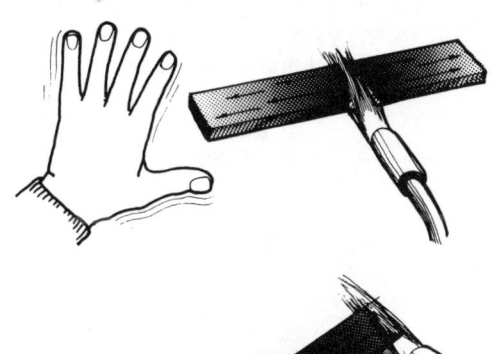

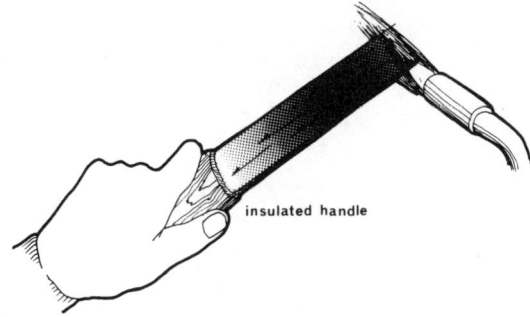

insulated handle

Fig. 4.3

Stripping knives, soldering irons, cooker or oven doors are made of metal and, when heated, quickly become too hot to be held or touched (Fig. 4.3). Handles are fitted to them, made from materials of low conductivity, such as wood or plastic. These materials resist heat passing through them and can be safely held a long time after the metals attached to them become very hot.

Table 5 shows common materials used in buildings. They are arranged in their correct order of THERMAL CONDUCTIVITY. Those at the top allow heat to pass through them quickly. Those at the bottom allow heat to pass through them much more slowly.

Table 5 **Thermal conductivity of building materials**

Copper – good conductor, poor insulator
Aluminium
Iron
Glass
Concrete
Brick
Water
Wood
Cork
Still air – poor conductor, good insulator

4.4 What is meant by the thermal conductivity of a material?

A measure of the speed at which heat travels through a material.

If a blowlamp flame is directed against a piece of steel not only does the area in front of the flame become hot but, in a short time, the entire piece of steel becomes hot. Heat travels through all materials but some will allow it to travel quickly while through others it may be very slow. This

GOOD CONDUCTOR	POOR CONDUCTOR

copper tipped iron

wooden handle

steel radiator

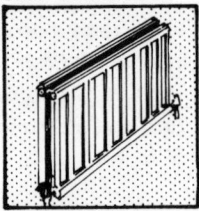

vacuum flask

POOR INSULATOR	GOOD INSULATOR

Fig. 4.4

heat movement is known as CONDUCTION. Materials which allow fast movement of heat are known as good conductors. When the movement is slow they are called poor conductors.

Copper is a very good conductor of heat and is called a material of high THERMAL CONDUCTIVITY (thermal means heat). In comparison, still air is a poor conductor so has a low thermal conductivity (Fig. 4.4).

The tip of a soldering iron is made of copper because of its high thermal conductivity. It takes the heat from the electrical heating coil and becomes hot very quickly.

The poor conductivity of still air is used when it is trapped in the cells of plastic foam. Foamed plastic is injected into cavity walls so that heat will travel very slowly through the trapped air and heat loss in the room is reduced.

Paint is more difficult to burn off steel than wood because of steel's high conductivity. Heat from the blow-lamp is absorbed by the steel and quickly passes through the metal before the paint becomes hot enough to soften.

4.5 Is the air hotter above a radiator than below it?

Yes, because hot air rises.

When air around a radiator is heated it expands and becomes lighter and less dense. This is how a hot air balloon rises (Fig. 4.5A). Cold air from other parts of the room moves into the space left by the warmed air. The cold air then in turn is heated and rises and so the process continues. As the warm air gets further away from the radiator it cools and contracts and sinks back to the lower levels, where it is heated again and rises. This circulation effect when air is heated is known as CONVECTION (Fig. 4.5B).

Convection fires are designed to allow cold air to come in at floor level, where it is quickly warmed by an electric or gas filament, and directed out into the room through a wide flue. The CONVECTION CURRENTS of air circulate until all the air in the room is warm, when the process gradually slows down. It does not stop, though, as the hot air is losing its heat to the walls and ceilings in the building all the time.

Fig. 4.5A

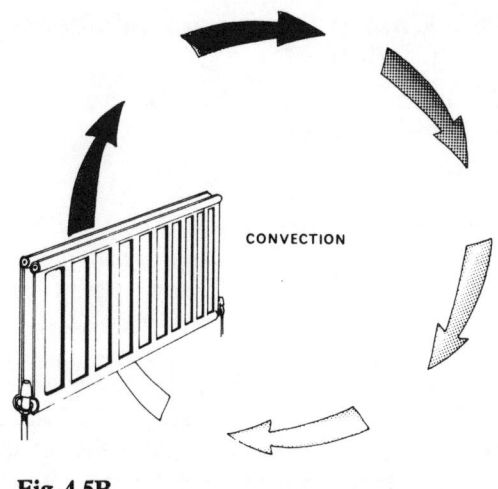

CONVECTION

Fig. 4.5B

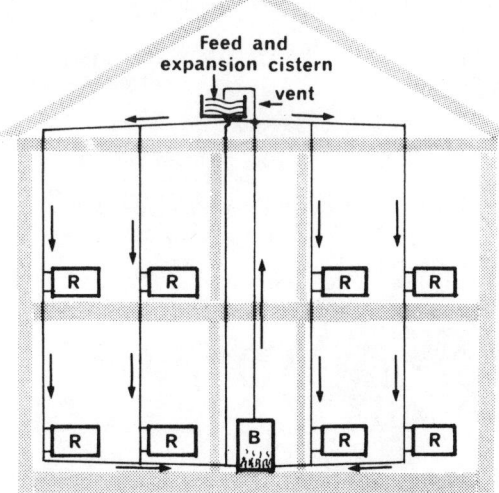

Simple gravity heating system

Fig. 4.6

4.6 How does a hot water radiator system work?

The heated water moves upwards through the pipes and radiators by CONVECTION.

Heat passes through water by convection in the same way that it passes through air.

When water is heated it becomes lighter and moves upwards or *floats* on the cold water.

In a hot water heating system (Fig. 4.6) the water is enclosed in pipes which run from the heater or boiler through the radiators and back to the boiler. When the water is heated it moves upwards pushing the cold water through the pipes. When the cold water reaches the boiler it becomes hot and moves upwards and keeps the hot water circulating. As the hot water loses its heat to the metal of the radiators and the air around them, it is pushed along by the hot water behind it back to the boiler for reheating.

Old radiator systems worked entirely by convection and needed very large pipes. The hot water moved very slowly and a lot of the heat was lost through the pipes before it reached some of the radiators which were a long way from the boiler. Modern systems are helped by PUMPS. A pumped system gets the hot water to move faster and makes sure that it gets to the furthest radiators. Also it allows smaller diameter pipes to be used.

4.7 How does heat get from the sun to us?

In the same way that a fire gives off heat, by RADIATION.

The sun sends out invisible heat rays called INFRA-RED RADIATION. They pass through space until they meet something which can absorb them. In this case our planet and its atmosphere (Fig. 4.7A).

Radiation often starts other forms of heat transfer. The outside of a house will become hot by radiation. This absorbed heat is passed through the walls by CONDUCTION to heat the inside of the house.

47

RADIATED HEAT FROM THE SUN

Fig. 4.7A

A person standing in front of a fire is quickly heated by radiation (Fig. 4.7B). The part of the body facing the fire will become hotter than the other side.

The air around the fire is heated by radiation also. More slowly the heated air rises and moves around the room by CONVECTION. Gradually the side of the person away from the fire becomes warm as the warmed air circulates around him.

Fig. 4.7B

4.8 What type of materials heat up slowly?

Usually thick or dense materials become hot more slowly than thin ones.

The rate at which RADIANT HEAT makes objects hotter depends upon their ability to absorb the heat and their thickness. Most thin materials will become hot quickly and allow heat to pass through them very fast. Most heavy, dense or thick materials become hot very slowly and allow heat to pass through them slowly. Materials which are good conductors of heat allow heat to pass through them quickly. The speed of the heating depends on the thickness of the material. A thin copper pipe will get hot quicker than a thick copper pipe.

The air in a building with a thin corrugated iron or asbestos cement roof will become hot much quicker by sun radiation than a similar building with a thick concrete roof.

A thin layer of plastic foam as a roof covering will let less heat through than a concrete roof of twice the thickness. The reason for this is the poor conductivity of the foam.

4.9 Is heat reflected?

Yes, from shiny, polished surfaces in the same way that light is REFLECTED.

Light and heat are both forms of energy and behave in similar ways. When light is shone on to a mirror it is reflected back. If the reflected light from a lamp can be directed on to one spot from a number of mirrors less light will be wasted from the bulb.

REFLECTIVE SURFACES also send back heat rays. Very little heat is absorbed or passed through a brightly polished or shiny surface. This is why an electric bar fire is made with a reflector

Fig. 4.9

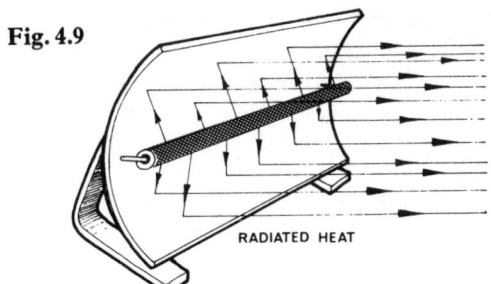

RADIATED HEAT

It is placed behind heating panels and radiators to reflect heat coming from the back of the panels into the room. It stops the walls absorbing it also.

Plaster board is available with aluminium foil fixed to one side to reduce the amount of heat that may pass through.

Highly polished or chromium plated steel is used behind electric or gas fires to reflect heat into the room.

(Fig. 4.9). More heat is directed forwards to where it is needed.

Bright metals are used to prevent heat loss in buildings.

Aluminium foil is the most common reflective surface used. It is put in oven linings to reduce the amount of heat which may be lost through the sides.

4.10 How can heat be contained in a building?

INSULATION will reduce the amount of heat which will escape through the walls and roof of a building (Fig. 4.10A).

Heat travels from one material to another. Cold

Fig. 4.10A

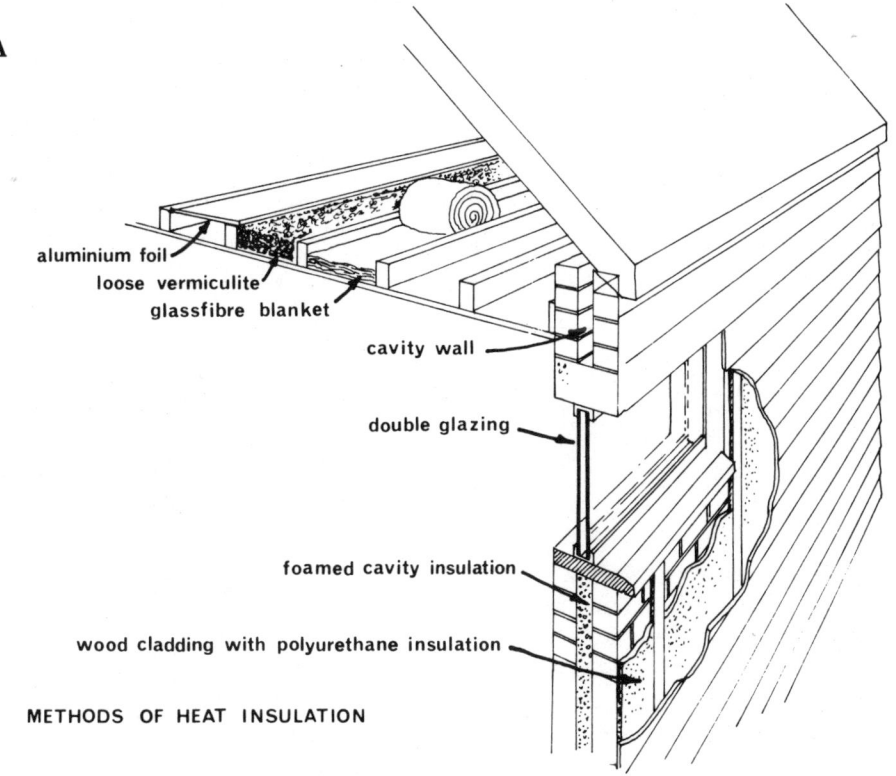

aluminium foil
loose vermiculite
glassfibre blanket
cavity wall
double glazing
foamed cavity insulation
wood cladding with polyurethane insulation

METHODS OF HEAT INSULATION

water will take in the heat from hot water and become warm. Cold air outside a house will absorb the heat from the warmer air inside the building. The heat will travel through glass, brick, concrete or wood, fast or slowly, to be absorbed by the cold air on the other side.

To reduce the amount of heat that may be lost through windows, walls or roofs insulating materials are fitted. These are poor CONDUCTORS of heat and form a good barrier between the warm air inside the building and the cold air outside.

Still air is one of the best insulators because it is a poor heat conductor. Most of the common insulating materials are traps for air. Glassfibre blankets are very light materials which hold a large quantity of air between their fibres. The blankets are laid between loft rafters or around pipes and reduce the amount of heat lost.

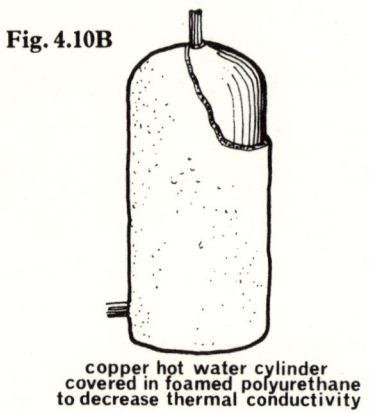

Fig. 4.10B

copper hot water cylinder
covered in foamed polyurethane
to decrease thermal conductivity

Expanded polystyrene and polyurethane are plastics which have been treated so that they expand into foams which trap air in the cells (Fig. 4.10B). They are used in the form of sheets or small pieces in between loft rafters, or injected into wall cavities to reduce heat loss. Vermiculite is used in small pieces and put into roof spaces or mixed with plaster to hold air in its cells.

Heat loss through windows can be reduced by trapping air either between two panes of glass (double glazing), or between a heavy curtain and the window. A heavily curtained, double glazed window is the best insulator.

4.11 Does a dark coloured surface become hotter than a light coloured one?

Yes, because dark colours absorb heat and light colours reflect it.

Light and heat are part of the same family of energy. Light can be seen, and heat is invisible, but they behave in the same way.

Surfaces react the same to heat as they do to light. A dark-coloured surface looks dark because it absorbs most of the light and reflects very little.

When a dark coloured surface absorbs most of the light rays it is absorbing heat also. If a light coloured surface reflects most of the light shone upon it, it also reflects most of the heat energy. So a black surface absorbs all the heat and becomes hot, and a white surface reflects the heat and remains cooler (Fig. 4.11).

Fig. 4.11

If a front door of a house facing south is painted in a dark colour and the frame in white, on a hot, sunny day the door may become so hot that it is painful to touch, whereas the frame will feel quite cool. In tropical countries the outsides of houses are painted white to help keep them cool.

4.12 Do all materials heat up and cool down at the same rate?

No, because not all materials require the same amount of heat to change their temperatures equally.

A material which requires very little heat to increase its temperature by a certain amount has a low SPECIFIC HEAT CAPACITY. These materials lose their heat quickly also.

Materials which require a lot of heat to raise their temperature have a high specific heat capacity. They also hold the heat for a long time.

Concrete has a high SHC and many electric storage heaters are lined with this type of material (Fig. 4.12). They heat up slowly during the night when electricity is cheap and give off the heat slowly during the day. Air has a low SHC. It is quickly warmed by a hot radiator but quickly cools when the radiator is turned off.

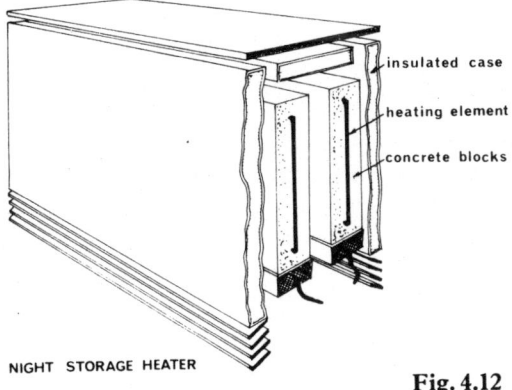

insulated case

heating element

concrete blocks

NIGHT STORAGE HEATER

Fig. 4.12

4.13 Is one fuel as good as another for producing heat?

No, it depends on their CALORIFIC VALUE.

Fuels which give off a lot of heat when burnt have a high calorific value. Those which give off less heat have a lower calorific value.

The old unit for measuring heat was the CALORIE. Some industries still use calories, and British Thermal Units (BTU). The modern unit for measuring heat is the JOULE. So a fuel which has a high *Joule rating* will produce much more heat than one with a low number. The calorific value of a fuel is the amount of heat, in Joules, produced when 1 kg of that fuel is completely burnt (Fig. 4.13).

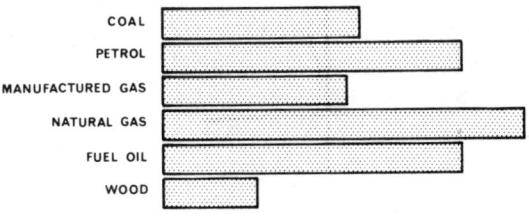

Comparison of heat given out by 1kg of fuel

Fig. 4.13

The amount of heat produced by burning one kilogramme of wood is less than that produced by burning one kilogramme of coal. Wood has a lower CV than coal.

Natural gas has a higher CV than town gas. So an oven burning natural gas will require less gas to heat the oven to 300°C than a similar oven using town gas.

Propane has a higher CV than butane. So a propane burner will produce more heat than a butane burner using the same amount of gas.

The body changes food into energy. Foods have different calorific values and people whose work requires them to use a lot of physical energy, or to train hard, or to keep warm in the cold, need to eat food having high calorific values. Potatoes produce more heat when eaten than apples and are called a high calorie food.

Food energy is now measured in MEGAJOULES (MJ).

4.14 What causes walls and ceilings above a radiator to get dirty?

Dust particles are carried by warm air onto any surface that is cooler than the air and they tend to stick to it.

A radiator heats the air around it, causing CONVECTION CURRENTS.

There is always dust in the air and this is moved around by the current of warm air.

As the convection currents move away from the radiator they get cooler and slow down.

Some of the heat from the warm air is absorbed into the walls and ceiling and some of the dust in the air is left on the surface.

The cool wall and ceiling areas nearest the radiator receive the warmest air and, therefore, the most dust (Fig. 4.14).

Fig. 4.14

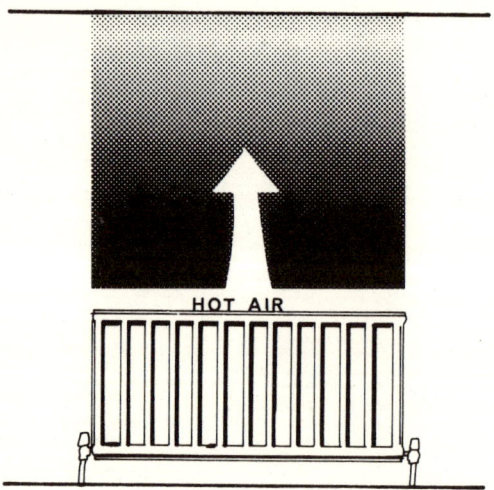

Dirty marks above radiators can be reduced by directing the convection currents away. The simplest way is to fit a shelf above the radiator. The dust will still be deposited but it will spread more evenly around the room and will not look so unsightly.

4.15 What is pattern staining?

Patches of dust, usually rectangular in shape or in lines, which may be seen on ceilings and internal wall surfaces.

Dust is always present in the air and it is moved about by warm air CONVECTION CURRENTS. As the heat from the warm air is absorbed into the cooler walls and ceilings the dust is left on the surface. The greater the difference in temperature between the warm air and the cool surface, the more heat will be lost and more dust will be seen on the surface. If the whole surface is not the same temperature, heat loss will vary and dust will stick at different rates. Most dust will stick to the cooler areas. This causes the patterns and they vary in shape according to how the building is constructed.

Most ceilings have a layer of plaster or plasterboard attached to timber joists. Since wood is a poorer conductor of heat than plaster or plasterboard, heat is absorbed faster through the colder space between the joists (Fig. 4.15). The difference in temperature between the two is usually small but sufficient to allow more dust to adhere to the cooler area between the joists than under them. These spaces will look dirtier. This problem can often be prevented by filling the space between the joists with an insulating material. The surface will then have a more even temperature. Dust will still stick to the surface but will be more evenly spread and no pattern will be seen.

If iron girders are used in place of timber joists

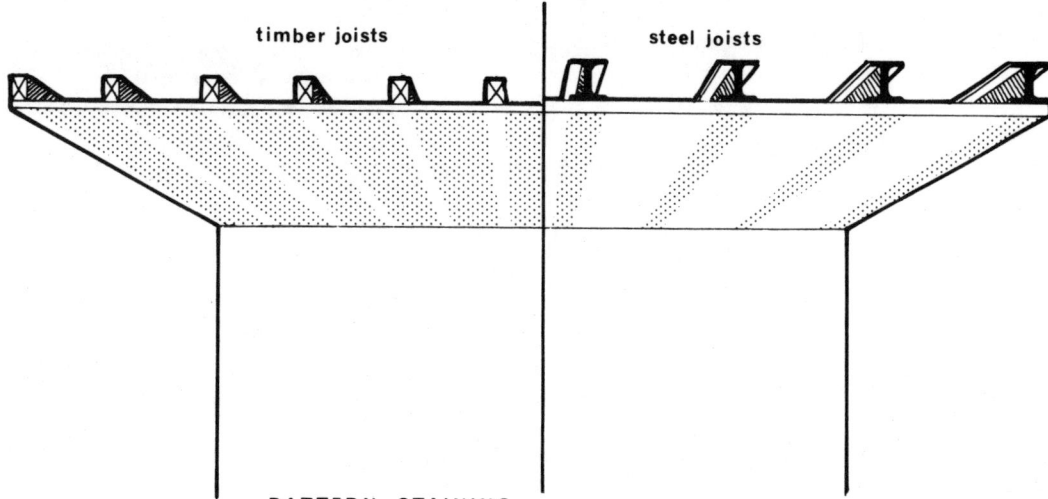

timber joists   steel joists

PATTERN STAINING

Fig. 4.15

a reverse pattern will occur. The surface between the girders will be warmer than under them because iron is a good conductor of heat.

A similar problem can sometimes be seen on walls made of building blocks. A block pattern shows even though the surface is plastered to a smooth finish. The blocks contain many air holes and are slightly warmer than the more dense cement and sand joints. This results in the joints absorbing more heat and more dust sticks to the surface in front of the joints than the blocks.

4.16   What is refrigeration?

Refrigeration is the transfer of heat from a substance which is to be cooled to somewhere else.

Heat flows freely, without assistance, from a hot substance to a cool substance. The hot substance gets cooler and the cool substance gets hotter. This can continue until both are of the same temperature. For example, heat inside a house flows through walls, windows and roof to the cooler air outside, making the house air colder.

Heat inside a radiator flows into a cooler room.

Refrigeration systems work by transferring heat from the inside of a refrigerator to the

Fig. 4.16

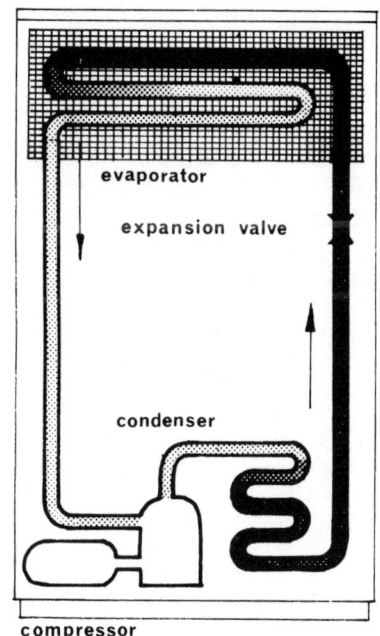

evaporator

expansion valve

condenser

compressor

53

outside. A gas is liquefied by compressing it and then allowing it to expand, which makes it cold. This cold gas, known as a REFRIGERANT, is circulated through thin walled pipes attached to the refrigerator (Fig. 4.16). The heat inside the refrigerator flows through the walls of the pipes and warms the gas. The warmed gas is piped to the outside of the refrigerator, where the heat is given off to the outside air by passing the gas through finned tubing. This heat can be felt rising from the back of a domestic refrigerator. The heat has been transferred from the inside of the refrigerator and so the foods in the fridge stay at a lower temperature than the outside.

4.17   What is a thermostat?

A piece of equipment that can cause a switch to be turned on or off as the temperature changes.

The most common type of THERMOSTAT is made from very thin strips of two different metals bonded together. This is a BIMETAL STRIP. When the strip becomes warm both metals expand, but one of them expands much faster than the other. The fast expanding metal cannot expand in a straight line because it is being restrained by the slower expanding metal. The only way it can increase its length is to curve. The metal that expands the most is on the outside of the curve, which is longer (Fig. 4.17).

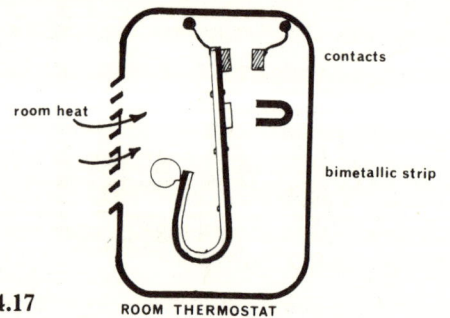

Fig. 4.17

room heat

contacts

bimetallic strip

ROOM THERMOSTAT

The bimetal strip is fitted into an electric circuit. One end is fixed, the other left free. When the strip is cool the free end is in electrical contact and the heater works. When the bimetal strip becomes warm it curves away from the contact and the heater goes off. As the strip cools it straightens again, the contact is made and the heater comes back on.

Thermostats are made to switch on or off electric, gas or oil heaters. They are put in gas and electric cookers; electric, gas or oil fired heating systems; electric irons and kettles; and refrigerators. They can be set to various temperatures. When the cooker, or the room air, or the iron or kettle element, has reached the set temperature the thermostat automatically switches off the heater. When the temperature drops the thermostat switches the heater back on. The switching on and off keeps the temperature at about the same level.

Car engines have thermostats also. These work in a different way. They contain a metal strip or a liquid which expands when hot and opens a valve, and contracts when cool to close the valve. It is placed in the cooling system to allow the engine to heat up quickly. The closed thermostat slows down the flow of water in the cooling system until the operating temperature is reached, when it opens to allow water to circulate and prevent the engine from getting too hot.

4.18   What is latent heat?

Latent heat is hidden heat.

If a kettle of water is boiled on a stove, and one thermometer is placed in the water in the kettle, and another thermometer very close to the spout, both read 100°C (Fig. 4.18). Water at 100°C is being turned into steam at 100°C. There is no rise in temperature, yet heat energy must be put into the water in the kettle to keep this process going.

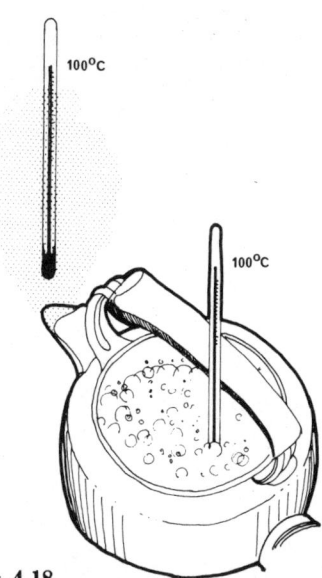

Fig. 4.18

The heat energy from the stove is being used to break the attraction between the water molecules and pull them away from each other to make a gas. This is where the energy from the stove is going. When steam at 100°C changes back into water at 100°C, this latent heat is given out. A burn from steam is far worse than a burn from boiling water.

If a drop of alcohol is poured on the hand it quickly disappears. It has turned into a vapour. The hand feels cold. Heat has been extracted from the hand and this energy has been used to vaporize the alcohol. In other words it has been used to pull the molecules of alcohol away from each other and much further apart.

When ice at 0°C changes to water at 0°C heat energy must be put in to do this. The heat energy is used to break apart the crystal structure of ice and does not raise the temperature. When water at 0°C freezes to ice at 0°C the heat energy is given out. This is why it feels slightly warmer after a fall of snow.

LATENT HEAT is the total heat absorbed or produced during a CHANGE OF STATE.

4.19 Is black a good colour to use on a roof?

Only if the sun's heat is required to be absorbed into the building.

Black surfaces absorb all the invisible infra-red energy sent out by the sun, as well as the visible light. A white surface reflects away most of the heat and will not become as hot as a dark surface. So when a maximum amount of heat is required to be absorbed surfaces should be black.

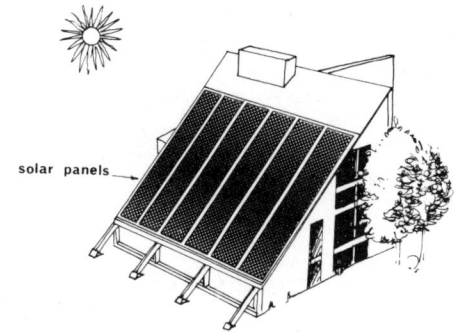

solar panels

Fig. 4.19

Solar panels are used in roofs of some buildings (Fig. 4.19). These absorb the sun's heat or SOLAR ENERGY and pass it on to heat the water in the building. The panels usually have a matt black finish to absorb as much heat as possible. A matt finish absorbs more heat than a glossy one because less light is reflected away. Even in winter, when the sun's heat is low, solar panels still absorb enough heat to raise the temperature of the water.

Many flat roofs are covered with asphalt or bitumen coated felt. These materials are naturally black and absorb a lot of solar energy. They are also THERMOPLASTIC so become soft when hot. Long exposure to hot sun causes bitumen felt to expand a lot. If the felt is not well adhered to the roof boarding it will expand into large blisters.

To reduce the amount of heat that asphalt and roofing felt absorb, white or light coloured granite chippings are put on the surface. These reflect away the sun's heat and prevent the materials becoming very soft or expanding excessively.

COMBUSTION

5.1 What is burning?

5.2 What is spontaneous combustion?

5.3 What causes a material to burn?

5.4 What is a combustible or flammable material?

5.5 Why is paper easier to ignite than wood?

5.6 How can a fire be put out?

5.7 What is the surface spread of flame?

5.8 How can the surface spread of flame be reduced?

5.1 What is burning?

Burning is a term which describes a fast chemical reaction between a substance and an OXIDISER (usually oxygen in the air). Normally a lot of heat is produced and often light. This is seen in the form of a glow and flames (Fig. 5.1A).

Fig. 51.A

Depending on what is burning there are many different things that happen, but in every case a chemical change takes place and the original material is not recoverable.

Gases are given off which can be poisonous. When coal burns the gas carbon dioxide is produced (Fig. 5.1B). This is the same gas that we breathe out. If there is not enough air, the gas carbon monoxide is made (Fig. 5.1C). This gas is highly poisonous. Carbon dioxide and some carbon monoxide are produced when petrol burns in a car engine.

Sometimes solid products remain. Coal, after burning, leaves an ash (Fig. 5.1D).

Burning is also called COMBUSTION.

Fig. 51.B

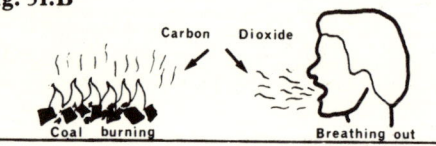

Fig. 51.C

Fig. 51.D

5.2 What is spontaneous combustion?

A fire which can occur without a flame being applied.

When certain materials are just put together a fire can start.

For example, if a pile of rags soaked in oil are stacked up, a fire may eventually start. What happens is that the oil slowly combines with the oxygen in the air. This chemical reaction produces a little heat which cannot be felt, but is enough to help more oil combine with more oxygen. More heat is given off and the same thing happens. The temperature in the pile of rags goes up and the reaction gets faster and faster. Eventually the oil is combining at such a rate that heat and flames can be seen. For this reason it is important that oil or solvent soaked rags must be laid out separately.

A similar reaction occurs with foam rollers. If after use in oil paint they are put in a closed container before being cleaned they can burn by spontaneous combustion.

5.3 What causes a material to burn?

A material will burn only if it is capable of burning, if it is made hot enough, and if there is plenty of oxygen available.

Not all materials can burn. Burning will occur only when three things happen.
(a) The material must be able to burn. It must be COMBUSTIBLE or FLAMMABLE.
(b) It must be hot enough. Combustible or flammable materials must have reached their IGNITION TEMPERATURE.
(c) There must be plenty of oxygen present. Burning is an OXIDATION process. The

material joins with oxygen at a very fast rate and gives off heat. Oxygen is obtained from the air.

These three conditions are sometimes called the triangle of combustion (Fig. 5.3). If one of them is removed or not available burning cannot happen.

Fig. 5.3

BURNING INCOMPLETE

5.4 What is a combustible or flammable material?

One which will burn.

COMBUSTION means burning. All building materials can be described as either those that will burn or those which will not burn. Those that burn are called either COMBUSTIBLE which are solid materials such as wood; or FLAMMABLE which are gases or liquids such as liquefied petroleum gas or petrol. Those that will not burn are called either NON-COMBUSTIBLE such as iron or concrete, or NON-FLAMMABLE such as air or water.

Fires occur because most combustible and all flammable materials give off a gas when heated which then ignites. The cause of ignition may be a flame, such as a match or gas pilot light; a spark

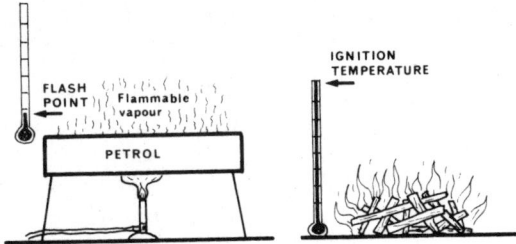

Fig. 5.4

from an electric motor or light; a spark from a clash of metals or metal on stone; or intense pressure, such as that caused in the combustion chamber of a diesel engine. Materials can burn by SPONTANEOUS COMBUSTION, when no spark or flame is needed to cause ignition.

The temperature at which materials give off the flammable gas is called either IGNITION TEMPERATURE, when referring to combustible materials, or FLASH POINT, for flammable materials (Fig. 5.4). The temperatures vary with the nature of the material. Petrol can have a flash point as low as −20°C, but wood may have to be heated to 200°C before it will ignite. Generally dense and thick materials will need to be heated to a very high temperature before they ignite. Thin materials, and those containing many holes, will ignite very quickly. Coal requires considerable heat before it will ignite. Wood will burn at a much lower temperature particularly if the wood is in small pieces because more oxygen can get to all its surfaces, also into its surface through the small holes in wood. For this reason wood is often used to start a coal fire, because it can be ignited more easily and, while burning, it generates enough heat to ignite the coal.

Wood dust is a most combustible material. Because the wood is powdered, much more surface area of wood is exposed to oxygen, and burning when it starts is very rapid.

5.5 Why is paper easier to ignite than wood?

A fire needs oxygen. The more oxygen that can get to the fuel the more readily and rapidly it will burn.

Paper contains many tiny holes into which the oxygen gets. Also, being thin, heat quickly passes through it and both sides will ignite. Wood also contains holes but not as many as paper, so less oxygen reaches its surface and ignition is slower. Wood is also thicker than paper, and heat from one burning side will take longer to pass through it to ignite the other side. Wood doors are accepted as fire resisting doors. A door of 50 mm thickness will resist a fire in a room for at least half an hour. After this time it will become sufficiently hot to burn right through.

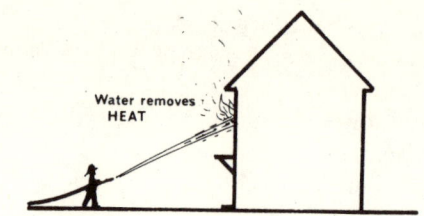

Fig. 5.6A

Fig. 5.6B

5.6 How can a fire be put out?

If any one of the three things which causes fire, heat, fuel or oxygen, is removed a fire will not continue to burn.

Doors and windows should be closed when a fire starts in a room, because once it has used up the oxygen in the room it will go out.

Fire extinguishers work in different ways. Some remove the heat, while others keep oxygen away from the fire.

Water cools the fuel and stops it burning. Burning is a chemical reaction which gives off heat. Water takes the heat away and the chemical reaction slows down and stops (Fig. 5.6A).

Gases such as CO_2 are heavy gases which lay around the fire and prevent oxygen reaching it (Fig. 5.6B).

Foam forms a blanket around the fire and isolates it from the oxygen and the burning stops. Powder also seals the fire from the oxygen.

Fig. 5.6C

Glassfibre blankets also stop the oxygen getting to the fire. They act a little slower than foam or powder because they trap air under the blanket which must be used up before the fire will go out (Fig. 5.6C).

5.7 What is the surface spread of flame?

The speed at which a fire spreads across a surface.

Some materials catch fire more easily and burn faster than others. The speed of burning depends on three factors.

(a) The surface must be made of a material that will burn. It must be COMBUSTIBLE.

(b) Some porous materials are less DENSE than other porous materials. They contain more oxygen inside them which means they will burn more quickly. Also, having large pores, more oxygen can get in to the burning surface.

(c) Large surface areas are in contact with a lot of oxygen around them so will burn faster.

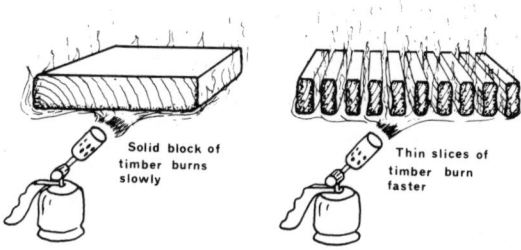

Fig. 5.7A

A solid block of timber will take much longer to burn than the same weight of wood cut into ten slices and spread out (Fig. 5.7A).

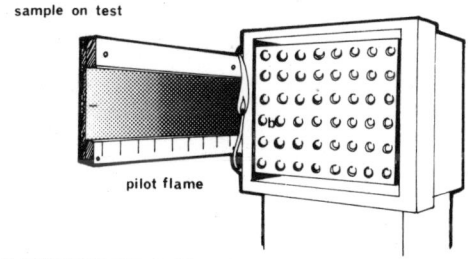

TEST EQUIPMENT FOR SURFACE SPREAD OF FLAME

Fig. 5.7B

All combustible building materials can be classified according to the speed at which they burn (Fig. 5.7B).

Fibre insulating board is porous, contains a lot of oxygen and burns rapidly. Some hardwoods are dense, contain a small amount of oxygen and burn more slowly.

5.8 How can the surface spread of flame be reduced?

By treating materials before fixing, or painting materials after fixing, with special coatings which slow down their rate of burning.

The special coatings are called FLAME RETARDANT COATINGS, and work in one or more of the following ways: they stop oxygen getting to the burning surface, they stop the surface getting hot, they give off gases which will not burn, or they stop the flammable gases escaping from the burning surface.

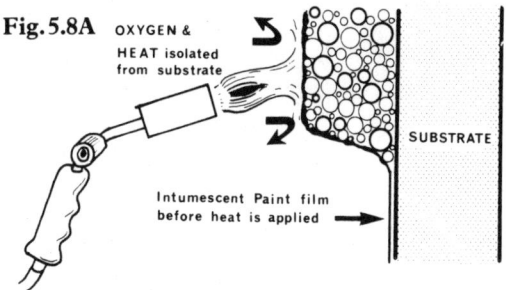

Fig. 5.8A   OXYGEN & HEAT isolated from substrate

SUBSTRATE

Intumescent Paint film before heat is applied →

INTUMESCENT coatings produce non-combustible gases when they get hot (Fig. 5.8A). These gases cause the paint film to bubble and form a layer of foam up to fifty times as thick as the original coating. This heavy layer of foam stops oxygen getting to the surface and insulates it from getting very hot.

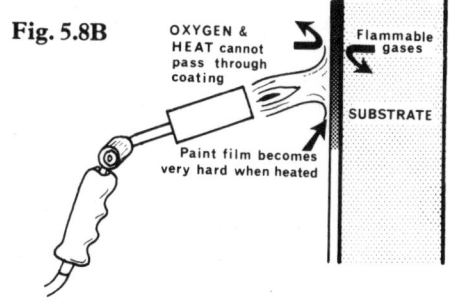

Fig. 5.8B   OXYGEN & HEAT cannot pass through coating

Flammable gases

SUBSTRATE

Paint film becomes very hard when heated

61

Another type of coating turns into a very hard, glasslike, heat resistant film when it gets hot (Fig. 5.8B). The flammable gases in the material cannot get through the film so they do not burn. The film also keeps oxygen away from the surface, again preventing combustion.

A third group of coatings gives off a non-flammable gas when it gets hot (Fig. 5.8C). This gas keeps the oxygen away from the surface and burning is prevented or slowed.

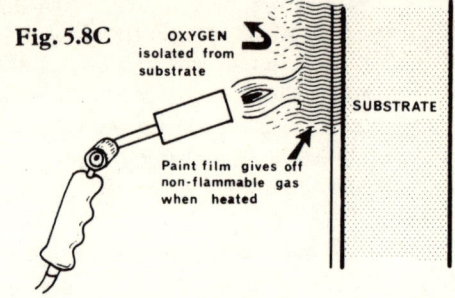

Fig. 5.8C   OXYGEN isolated from substrate

SUBSTRATE

Paint film gives off non-flammable gas when heated

SECTION SIX

ENERGY

6.1 Why do materials get hot when they are rubbed together?

6.2 How can friction be made to work for us?

6.3 How can friction be reduced?

6.4 How does a petrol engine work?

6.5 How does a diesel engine work?

6.6 What is static electricity?

6.7 Does electricity pass through some materials more easily than others?

6.8 What causes an electric cable to get hot?

6.1 Why do materials get hot when they are rubbed together?

Because of FRICTION which is the resistance of one material being rubbed against another.

When any materials are rubbed together heat is produced and they become hot (Fig. 6.1A). The harder or longer they are rubbed the hotter they get. Some materials can be rubbed against each other with very little effort, like a soft duster over a hard, shiny, polished surface. Friction here is very slight. There are some materials which require a lot of effort to rub together, like a file over rusty steel. This is because friction is very great and the file will get very hot quickly.

Fig. 6.1A

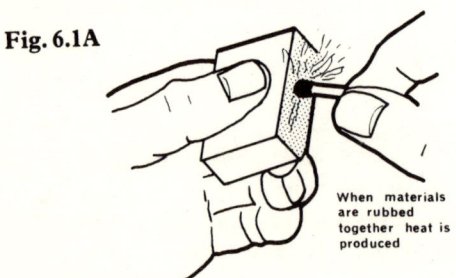

When materials are rubbed together heat is produced

Fig. 6.1B

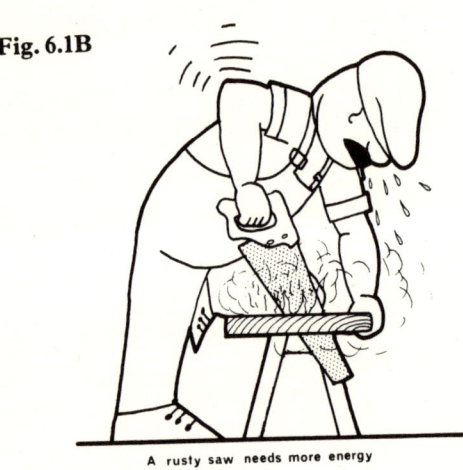

A rusty saw needs more energy and becomes hot

A rusty saw requires more energy to push it through a piece of wood than a clean shiny saw (Fig. 6.1B). So it will become hot. Also the amount of energy needed to push a saw through a piece of wood is greater than that needed to saw through expanded polystyrene. Therefore the wood saw will become much hotter than the polystyrene cutting saw.

6.2 How can friction be made to work for us?

There are many day to day things, as well as building processes, which are possible by the use of FRICTION.

The soles of plimsolls are soft and textured so that they grip the floor. They are designed to produce a higher friction between the plimsolls and the floor. So when a person wearing plimsolls runs on a smooth floor there is a resistance to the two surfaces slipping easily over each other (Fig. 6.2A). A hard, smooth soled shoe will slip because there is very little friction between the two surfaces.

Fig. 6.2A

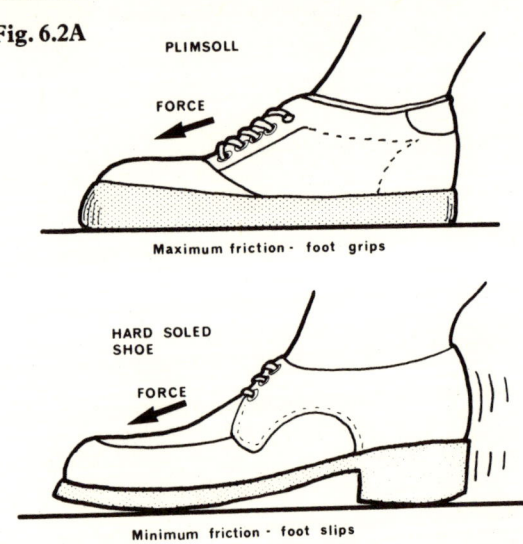

PLIMSOLL

FORCE

Maximum friction - foot grips

HARD SOLED SHOE

FORCE

Minimum friction - foot slips

Fig. 6.2B

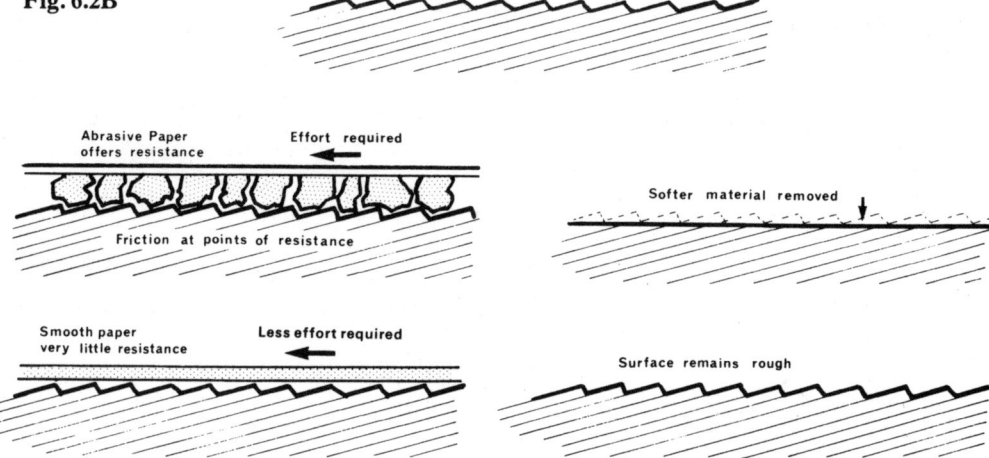

Rough texture of timber

Abrasive Paper offers resistance

Effort required

Friction at points of resistance

Softer material removed

Smooth paper very little resistance

Less effort required

Surface remains rough

Timber surfaces resist being rubbed with glasspaper. Friction causes the very rough texture of the glass to grip the less rough wood and the softer material is abraded. The smooth back of the glasspaper would pass over the timber easily but no abrasion would occur (Fig. 6.2B).

Brake linings are made of high friction materials. This makes sure that when they contact the turning wheel they do not slip. As they grip the wheel movement is slowed and the linings are slowly worn down. They also get hot.

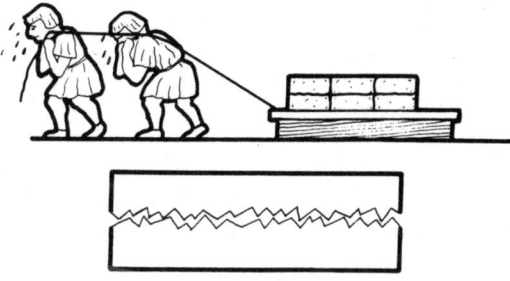

Fig. 6.3A

6.3 How can friction be reduced?

By keeping surfaces in as little contact with each other as possible. Or by using a LUBRICANT between the two surfaces so that they are not in contact at all.

A wheel is one example of reducing the amount of contact between surfaces. A load of bricks requires less effort to move if it is put into a wheelbarrow rather than dragged along on a palette. The wheel of the barrow will turn more

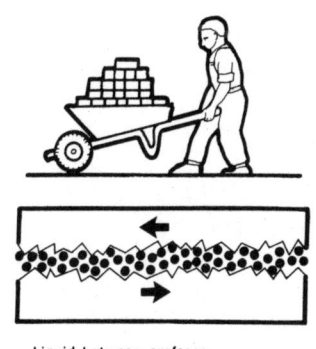

Liquid between surfaces

Fig. 6.3B

easily if the axle moves within a ballbearing rather than turning around a solid bar. Wheels and ballbearings have a smaller area in contact with the other surfaces so there is less resistance and less work is needed to move them (Fig. 6.3A).

Any liquid used between two rubbing surfaces will allow them to move easier (Fig. 6.3B). Oil allows engines to run at great speeds because it forms a film between the moving surfaces. If there was no oil between the moving surfaces the friction would produce so much heat that the metals would weld themselves together – the engine would seize up.

A wet abrasive paper becomes less hot than a piece of dry glasspaper when used for rubbing a painted surface because water acts as a lubricant. Much more abrasion is carried out with less effort.

Airless spray hoses are coated on the inside with special plastic coating so that the paint being pushed through at very high pressures passes smoothly along the hose with the minimum of resistance.

6.4   How does a petrol engine work?

A petrol engine works by using the energy caused by igniting the highly flammable mixture of petrol vapour and air within a very small space.

A petrol engine contains a number of cylinders in which pistons continuously move up and down. As each piston moves down to the bottom of its stroke a mixture of petrol vapour and air is forced in to fill the space. The piston then rises and compresses this mixture into a small volume at which point it is ignited by a spark from a plug. As it burns, hot expanding gases are formed. The part which can be moved to make space for this large amount of expanding gas is the piston, and this is pushed down by the gas pressure. The next time up, the piston pushes the burnt gases into

the exhaust system and the process starts again. This is a four stroke engine (Fig. 6.4).

The energy created by the piston movement is directed by a crankshaft and a series of gears to perhaps turn the wheels of a car or make an air compressor compress air.

On small engines, like those used for mopeds or air compressors, only one cylinder may be used. These are often two stroke engines. For larger engines a number of cylinders are used, each burning their gases in a different order to supply a constant force of energy.

Engines which burn fuels inside a cylinder are called INTERNAL COMBUSTION engines.

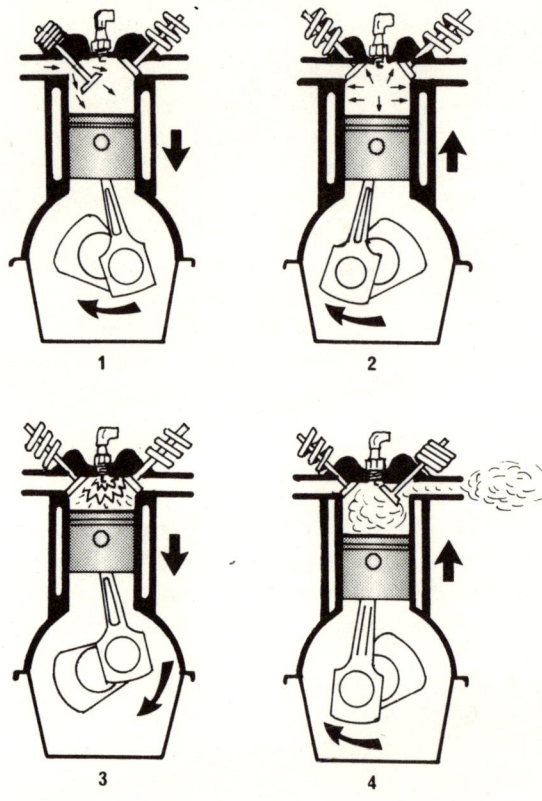

Fig. 6.4

6.5   How does a diesel engine work?

When air is rapidly compressed into a tiny space it becomes very hot. If a vapour which is flammable when mixed with air is sprayed into the compressed air when it is hottest it will ignite. When this happens the mixture burns and a large amount of expanding gas is made.

A diesel engine has a number of cylinders within which pistons move up and down. As they move down air rushes in to fill the space in the cylinder.

As they move up the air is compressed into a very tiny space and is made very hot. At this point diesel oil is sprayed into it and ignites. As it burns and expands the piston is driven down very fast (Fig. 6.5).

Each cylinder is timed to ignite in quick succession and the energy made by the fast moving pistons is passed on to move a vehicle or operate an air compressor.

A diesel engine does not need a sparking plug.

Diesel oil is a low grade petroleum fuel.

6.6   What is static electricity?

An electric charge caused by FRICTION.

All matter is made of ATOMS. The atoms are made of two parts, an inner and an outer part. The outer part consists of small pieces of NEGATIVE ELECTRICITY, called ELECTRONS When certain materials are rubbed on other materials friction causes some of the electrons to leave one material and to stick on the other. One material now has more electrons than the other. It is negatively charged with STATIC ELECTRICITY. If the material is handled the excess electrons flow through the fingers to earth. The other material is positively charged. If this material is touched, electrons flow from earth to replace the missing ones. Either way, an electric current flows, which is STATIC ELECTRICITY.

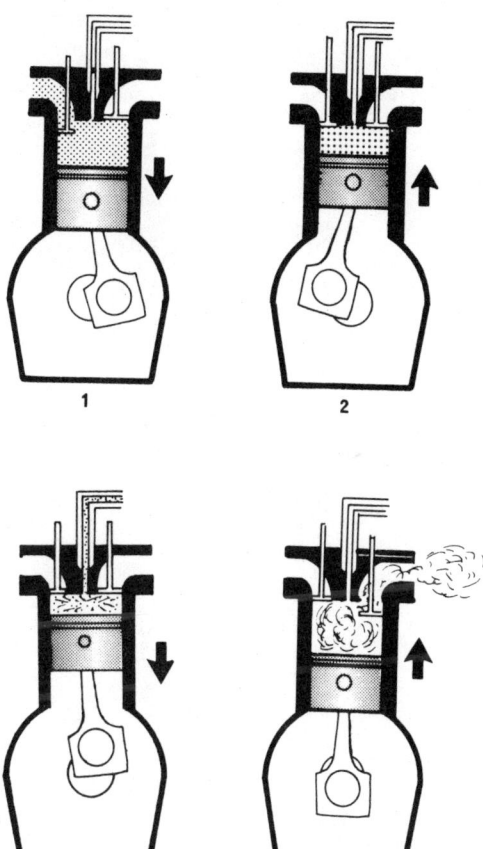

Fig. 6.5

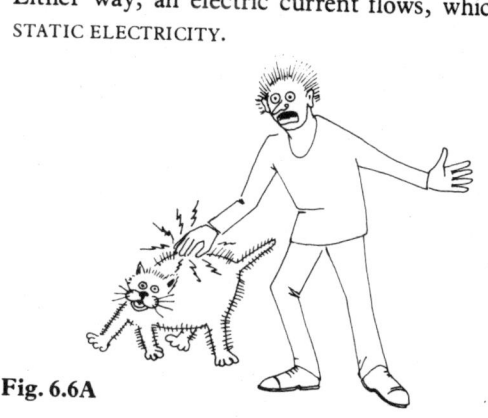

Fig. 6.6A

Very high voltage can be reached with static electricity, often many thousands of VOLTS. However, because the amount of electricity that actually flows is very small the static electricity that is occasionally felt in everyday life normally does no harm. Unless of course it is a bolt of lightning!

People experience the effects of static electricity, particularly on a dry day.

When a cat's fur is stroked small sparks of electricity can be heard (Fig. 6.6A). Walking across a carpet made of synthetic material and then touching a metal door handle sometimes gives a person a slight shock. Sliding out of a car with a plastic seat and touching the door often causes a small spark and shock.

Taking off a nylon shirt by drawing it over the body often causes electric sparks if the room is dry.

At night static electricity sparks can be seen.

Just as a car picks up a charge of static electricity by the friction of the rubber tyres on a dry road, so an aircraft may pick up a charge by the friction of the air flowing by. It can store this electricity and remain charged for a long time. For this reason an aircraft is earthed safely before refuelling, otherwise a spark may jump to the fuel line and cause a fire.

A fluid hose of an airless spray unit may produce a static electrical charge by the friction of the paint against the hose wall (Fig. 6.6B). When flammable materials are being sprayed in a closed area the hose must be earthed to prevent a spark causing the solvent vapour to ignite.

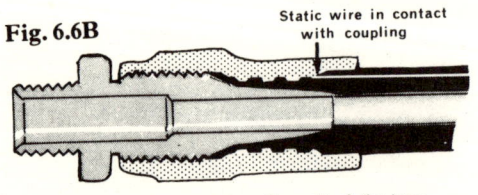

Fig. 6.6B

Static wire in contact with coupling

Friction of the paint against the wall of the hose produces static electricity.

6.7 Does electricity pass through some materials more easily than through others?

Yes. Materials known as CONDUCTORS allow electricity to flow along them very easily. Others resist the flow of electricity. These are called INSULATORS.

The best conductors are metals. Silver is the finest of all but for commercial purposes copper and steel are used to make wires, cables and fuses.

Poor conductors or insulators are used to cover and protect wires and cables thus preventing a short or an electric shock. Those in common use are rubber and plastics, such as P.V.C. (Fig. 6.7A).

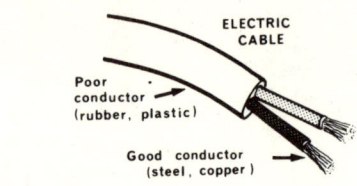

ELECTRIC CABLE

Poor conductor (rubber, plastic)

Good conductor (steel, copper)

Fig. 6.7A

Other insulators, such as glass and ceramics, are used to support wires and cables, such as those on electricity pylons or the overhead cables on electric trains (Fig. 6.7B).

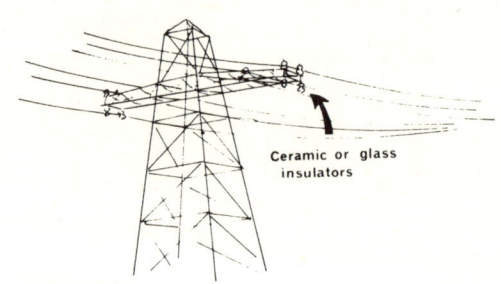

Ceramic or glass insulators

Fig. 6.7B

6.8 What causes an electric cable to get hot?

Passing too much electricity through it for its rating, or using a faulty cable.

Electricity flowing in a wire makes the wire heat up. Obviously this rise in temperature must be kept as small as possible. Normally it cannot be felt.

The amount of electricity flowing along a wire is measured in AMPERES, or AMPS for short. Electric cables, fuses, plugs and sockets are rated according to the amount of electricity they will safely carry. This is always shown in number of amps.

The more current that must pass along a wire, the thicker the wire must be in order to carry that current safely without a large rise in temperature.

A cable with a rating of 3 amps will be quite suitable for a light bulb, record player or a small power drill, which require 1.5 amps. If the same cable was to be used on a compressor with a rating of 18 amps it would gradually become hot. It could become hot enough for the cable's insulated cover to burn, and perhaps start a fire. The cable could melt also. This happens because the compressor is taking more electricity from the supply than the cable was made to carry. The cable is not thick enough to carry 18 amps safely without heating up.

A similar problem can occur when a cable is kinked or used in a tight roll. A kinked cable may have caused a wire to break and so less wire is available to carry the electricity safely. In a roll of wire some of the heat given off is passed to other parts of the wire. The heat builds up and the insulation melts and causes a *short*. Too much current flows then and if the fuse does not blow a fire may start (Fig. 6.8).

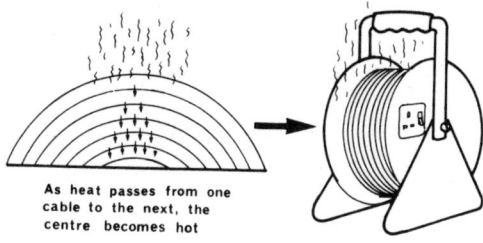

As heat passes from one cable to the next, the centre becomes hot

Fig. 6.8

COATINGS

7.1 How does a paint film expand and contract?

7.2 How do paints dry?

7.3 What is the difference between a catalyst and a hardener?

7.4 What is viscosity?

7.5 How is viscosity measured?

7.6 What effect does heat have on plastics?

7.7 What is a solvent?

7.8 What is thixotropy?

7.9 How does a paint remover work?

7.10 How do emulsion films allow moisture to pass through them?

7.11 Why do some paints require regular stirring?

7.1 How does a paint film expand and contract?

By moving with the surface on to which it is applied like a layer of elastic. This property of paint is known as FLEXIBILITY or ELASTICITY.

When paints dry the small particles of resins and oils all knit together to form a film like a piece of woven fabric with the pigment trapped between them. Some resins are brittle and produce paint films which break easily. Others are flexible and produce films which are able to bend and stretch.

There are a few paint films which will only bend and stretch to a very limited extent. The most common of these are spirit based paints and varnishes based on shellac. All paints become brittle with age and old paint films will eventually crack if the surface on to which they are applied expands or contracts.

Other paints, particularly brushing type gloss paints, are very flexible and will stand a considerable amount of movement.

A surface can move back and forwards just as a thin piece of metal can be bent. Or it can stretch or shrink like timber does when it becomes wet and dries out. Or it can expand or contract like metals when they become hot and cold.

Most gloss paints have sufficient elasticity to stand these movements. For example, an aluminium up-and-over garage door which often bends; normal softwood windows or doors which move with the changing climate; and radiators which expand and contract. Paints with less resin or oil and a larger amount of pigments, like spirit paints or eggshell finishes, are not so flexible and may crack.

7.2 How do paints dry?

Either by a simple process of the solvent evaporating leaving the solid material as a film, or by a more complicated process of the medium chemically changing its form from a liquid to a solid.

There are four important ways by which paints dry, depending on the type of medium used.

EVAPORATION is a simple drying process which only involves a change in condition. When the paint is applied the solvent evaporates from the film leaving the resins as a dry film. No CHEMICAL CHANGE takes place.

Shellac knotting, bitumen and cellulose are all manufactured by dissolving the resins in a suitable solvent. This forms a SOLUTION of resin and solvent. When the paints are applied, the solvents evaporate leaving the resin in a continuous film over the surface (Fig. 7.2A).

If the paint's original solvent is poured on to the dried film, the film softens and changes back to a wet film. As the solvent evaporates the paint dries again.

Because the dried film can be easily reversed back to a liquid these finishes are known as REVERSIBLE COATINGS.

OXIDATION is a more complicated drying process in which oxygen in the air combines by a chemical reaction with the drying oils and resins (Fig. 7.2B) The particles of oil and resin join together, helped by the oxygen, to form larger particles and eventually become a solid dry film. Once the paint or varnish is dry the process cannot be reversed. They cannot be softened by their own solvents and are known as NON-REVERSIBLE coatings. This method of drying applies to AIR DRYING primers, undercoats and finishes which contain drying oils and resins.

POLYMERIZATION is a drying process of some resins used in paints and adhesives which will not dry by absorbing oxygen from the air or by evaporation of the solvents (Fig. 7.2C). The resins have to be mixed just before use with a special material called a hardener or accelerator.

72

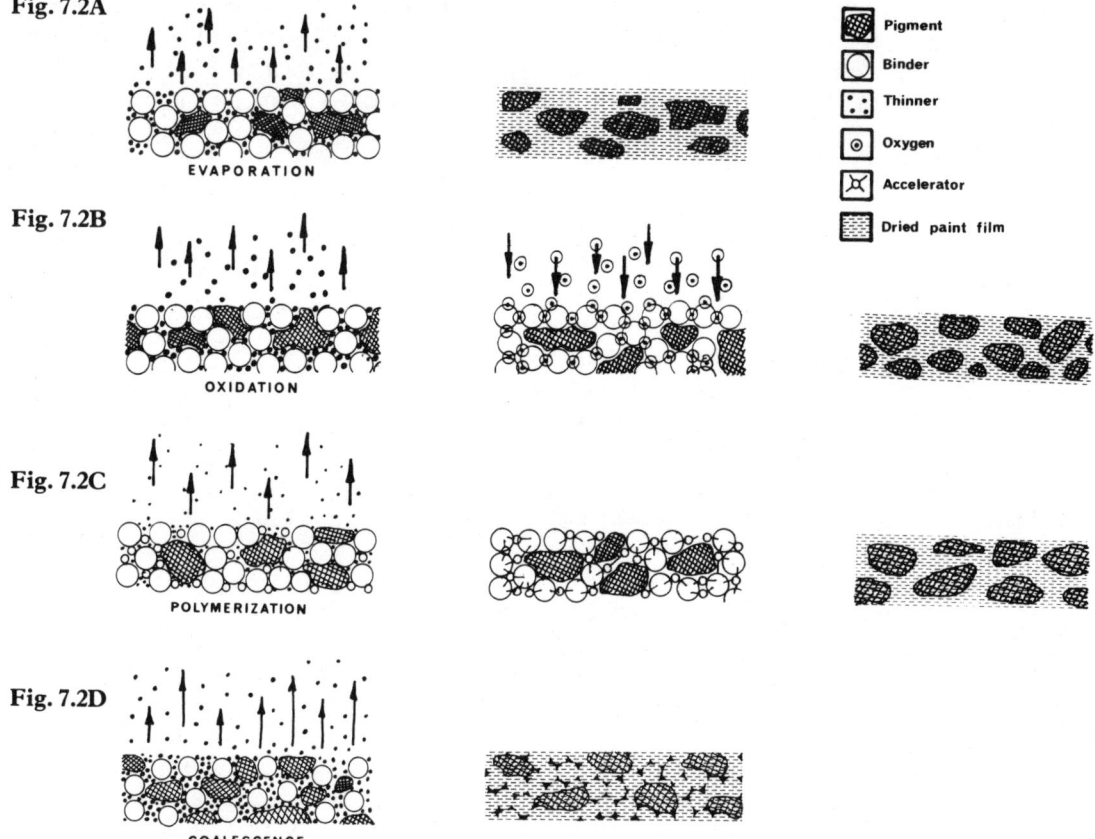

Fig. 7.2A

EVAPORATION

Fig. 7.2B

OXIDATION

Fig. 7.2C

POLYMERIZATION

Fig. 7.2D

COALESCENCE

▨	Pigment
○	Binder
⬚	Thinner
⊙	Oxygen
⊠	Accelerator
▦	Dried paint film

When the two materials are mixed together a chemical reaction takes place and the particles of resin join together to form larger particles and eventually become a hard non-reversible film. This process is known as CURING.

Once the chemical reaction has started it cannot be stopped, although low temperatures will slow down the reaction and heat will speed it up.

The most commonly used materials which dry by this process are two pack polyurethanes, two pack epoxies and polyester resins used in glass reinforced plastics.

COALESCENCE is the way in which paints and adhesives made by dispersing very small particles of resin in water dry (Fig. 7.2D). After application the water evaporates and the particles of resin join together to form a dry film. The spaces left by the water are not fully closed and this results in a fine honeycomb paint film. This is a characteristic of PVA and acrylic emulsion paints and adhesives and water thinned primer/undercoats.

7.3 What is the difference between a catalyst and a hardener?

A catalyst is a substance which speeds up a

73

chemical reaction, but remains itself unchanged at the end of the reaction.

A hardener, or accelerator, reacts with a resin and becomes part of a chemical change which converts both the resin and hardener into a hard film.

Two pack materials will not dry or CURE until the two components are mixed together. The materials in each pack are carefully chosen to react with each other to cause a CHEMICAL CHANGE to take place. The particles of resin and hardener link together, converting the liquids into a solid resin. Once mixed a whole tin of resin and hardener will turn into a solid in a few hours.

The most common types of two pack materials are based on polyurethane resins which use isocyanate hardeners, and epoxy resins which use amine or amine adduct hardeners. When the two are mixed in correct proportions they react, link up with each other and change into a very hard resistant film.

Unlike hardeners, true CATALYSTS cause components in a reaction to work faster but are not themselves an essential part of the resin film.

Driers used in oil paints are a form of catalyst because they help the oil absorb more oxygen quickly.

Catalysts are used also to speed up the reaction of peroxides which are used as accelerators in polyester resins and glass reinforced plastics.

the molecules of the fluid, which are forces of FRICTION. Oil molecules hold onto each other much more tightly than do water molecules. So an oily liquid takes a long time to move. Oil and varnish have a high viscosity.

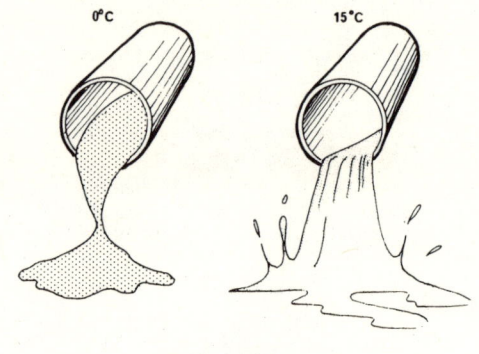

OIL PAINT

Fig. 7.4

When the temperature rises, oil gets easier to pour (Fig. 7.4). Its viscosity goes down. Engine oil becomes thinner as an engine gets hotter.

In freezing weather it is sometimes hard to get a car into gear as the oil in the gearbox has increased its viscosity. This property of oil is not desirable, so in modern cars oil additives are put in so that the viscosity stays the same over a wide range of temperatures.

A gloss paint or varnish used on a cold day is thick and difficult to apply. Used on a hot day the same material is quite thin and easy to apply.

7.4 What is viscosity?

The property of a liquid of resisting flow.

Pouring half a litre of oil into an engine takes longer than pouring the same amount of water down the sink. Pouring varnish is even slower. The speed at which a liquid flows is caused by its VISCOSITY. Viscosity is caused by forces between

7.5 How is viscosity measured?

By measuring the time that a standard amount of liquid takes to pour from a container.

It is possible to measure the difference in the VISCOSITY of various liquids by checking their resistance to flow. Paint manufacturers check the viscosity of every batch of paint made to make sure that they are of similar quality.

A practical method of measuring viscosity is to time a liquid passing through a hole in the bottom of a cup (Fig. 7.5). The equipment used for this is called a British Standard B4 flow cup viscometer. The cup is filled with the liquid to be measured and the time it takes to empty is checked with a stop watch. Thin paints, such as etch primers, will pass through in about 20 seconds. Thicker paints, such as gloss finishes, can take several minutes.

The temperature of a material is important when measuring its viscosity. Cold materials take longer to flow than warm materials.

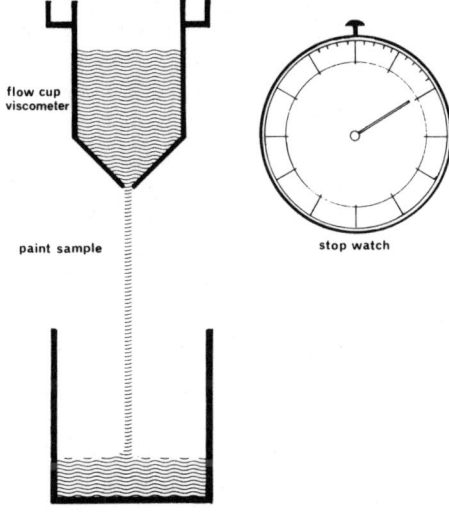

Fig. 7.5

Manufacturers of cellulose and stoving paints state very clearly what viscosity their materials should be before they are sprayed. They will stipulate a flow between 18 and 24 seconds at a certain temperature. This is usually 15°C. If a paint's viscosity is too thick it will take more than 24 seconds to flow out of the cup. By gradually adding more thinners the viscosity of the material is lowered until its flow takes less than 24 seconds but more than 18.

7.6 What effect does heat have on plastics?

It causes some plastics to soften and melt and other types to harden. If sufficient heat is applied plastics will burn.

Plastics can be divided into two main groups: those which soften with heat and are called thermoplastic, and those which harden with heat and are called thermosetting.

THERMOPLASTICS soften when heated above 80°C. Although they become very soft and pliable they change back to their normal hardness when cooled. Plastics in this group include polyvinyl choride (PVC) and acrylics. This property of plastics is used to form shapes like acrylic bowls and PVC signs. Bitumen is also a thermoplastic material. It is made hot and spread onto roofs as asphalt. When it becomes hot from the heat of the sun it becomes soft again. When cool it becomes very hard.

Heat causes other types of plastics to set or harden and once they are CURED heat has no further effect on them. These are known as THERMOSETTING plastics and include materials such as polyester resins used in fibreglass and Bakelite. Stoving enamels are thermosetting.

7.7 What is a solvent?

A liquid in which something can be dissolved or be floated.

The word SOLVENT is used in a general way for liquids which work in three ways.

They make a SOLUTION (Fig. 7.7A). Sugar dissolved in water, or shellac resin dissolved in methylated spirits are solutions. The sugar and resin molecules are separated and dispersed by the solvent and completely dissolved. Sugar or resin is the SOLUTE. Water or methylated spirit is the SOLVENT.

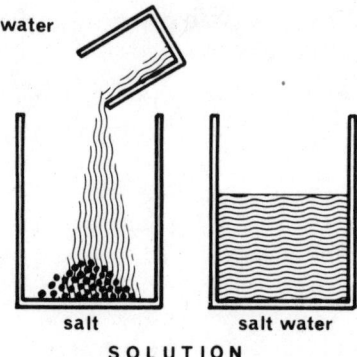

SOLUTION

Fig. 7.7A

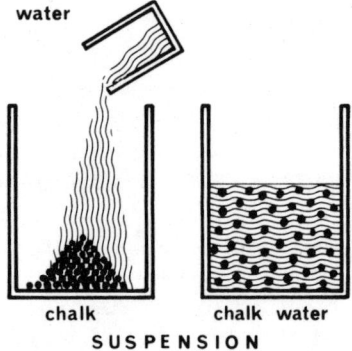

SUSPENSION

Fig. 7.7B

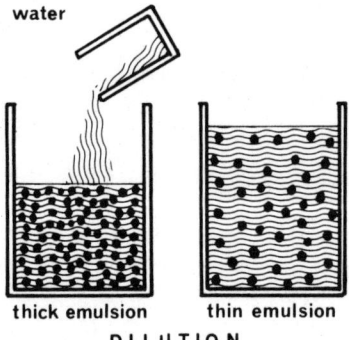

DILUTION

Fig. 7.7C

They make SUSPENSIONS (Fig. 7.7B). Some materials will not dissolve in a liquid but the particles remain floating in the liquid. Chalk is not dissolved in water. The particles can be seen floating in the water. Oil paints are suspensions of pigments in a solution of oil and resins.

They also DILUTE or *thin* other liquids (Fig. 7.7C). Some cleaning chemicals are supplied in a paste form and have to be diluted before use with a solvent (or DILUENT) like water. Emulsion paint is sometimes too thick to use and its VISCOSITY has to be reduced with water.

Other things which are necessary to know about solvents are as follows.

SOLVENT POWER or how good the solvent is at dissolving or thinning. Some solvents are more powerful than others. For example, cellulose paints require only a small amount of a powerful solvent such as amyl acetate to be thinned. The same paint will require a large amount of a weaker solvent to thin it the same amount.

EVAPORATION RATE or how long does the solvent take to dry. Solvents in paints, varnishes and adhesives are used only to make the materials easy to apply. Once applied the solvent should evaporate completely and quickly. Some take a few seconds, like cellulose paints. Others may take an hour.

PURITY, or does it contain grease and dirt. Impure solvents can slow the drying of paints or adhesives. Commercial paraffin is not as pure as white spirit and if used to thin oil paints can slow down or stop them from drying.

SUITABILITY, or is it the right solvent for the job. When paints, varnishes and adhesives are made solvents are used which mix easily with the material. Water will not mix with oil paints, and white spirit will not mix with emulsion paints.

7.8 What is thixotropy?

The property of a material which has a jelly-like state when still, becomes liquid when stirred or shaken, and returns to a jelly-like state when left to stand.

Most paints, particularly emulsion paints, are partially thixotropic. They are not true solutions. True solutions are those in which a material has been broken up into very tiny particles by a solvent and travels freely in the liquid. Paint mediums are not true solutions because the particle size of the oil and/or resins in the solvent is larger and they do not move so freely. This is a COLLOIDAL SOLUTION. Starch, glue, rubber and gelatine are examples of COLLOIDS. Because of their slow movement they interlock readily and restrict flow. They become rigid or GEL.

Many paints and adhesives are treated during manufacture to make this colloidal state greater.

One way in which materials are made thixotropic is by making the particles shaped like dumb-bells. They are like very small magnets having North and South ends which attract each other forming a rigid pattern. When an attempt is made to move these particles they resist by bouncing back like a jelly. The particles can only be separated by stirring or shaking vigorously. This causes them to move around more freely with less resistance like an ordinary liquid. When the stirring or agitation stops, the particles are able to attach themselves to each other again and the material returns to a thick jelly-like state.

Thixotropy is most useful in paints and adhesives because thick layers can be applied in one application and as the material sets into a gel quickly, runs and sags are less likely to occur. Gelled materials are more easily applied to vertical surfaces because they have greater resistance to flowing downwards.

7.9 How does a paint remover work?

Spirit paint removers contain solvents, or spirits, which are capable of dissolving the hardened oils and resins which form a dried film of paint. Caustic soda or chemical paint removers saponify hardened oils.

There are many solvents which are added to paints and varnishes. They make the paint less viscous, or thinner, so that they can be easily applied. They work by dissolving the liquid oils and resins in the material. It is usually found that a solvent which thins a liquid paint *will not* dissolve the paint after it has dried. This is because when a paint dries it changes into a new material which is unaffected by the solvent.

A few paints, which are reversible coatings, are soluble in their own thinner after they have dried. For example, acetone thins and dissolves both liquid and hardened cellulose paint, so can be used to remove the dry paint. Bitumen paints can be removed with their own thinner also.

Some solvents have the power to dissolve a large range of hardened oils and resins of dried paints. It is these solvents which form the basis of liquid spirit paint removers.

As the dry paint film is dissolved it absorbs the solvent and swells. As it swells the film needs more room and the only way it can get it is to wrinkle. This is the reason why paint films separate from the surface and seem to bubble up when they are attacked by spirit paint removers.

Caustic paint removers work by changing the oils in the paint film into a soap. It is a process known as SAPONIFICATION. The sticky, slimy soap can be dissolved in water.

7.10 How do emulsion films allow moisture to pass through them?

Moisture, in the form of vapour, passes through the film, which is like a fine honeycomb.

Emulsion paints are DISPERSIONS of very small particles of resin suspended in water. When the paint is applied the water evaporates and the particles of resin stick together to form a layer. As the film dries not all the spaces left by the water are filled up by the resin. Some remain open leaving the film as a very fine honeycomb (Fig. 7.10).

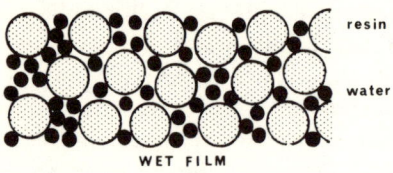

WET FILM

resin

water

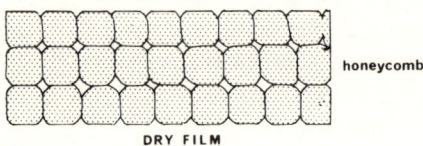

honeycomb

DRY FILM

Fig. 7.10

The spaces between the particles of resin are very small but it is possible for moisture, in the form of vapour, to pass through the film. It will not allow a large amount of moisture to pass through it. Emulsion paints can be used on surfaces such as new plaster. They allow the surface to continue to slowly dry out after it has been decorated.

However, if the surfaces are very wet the water cannot escape and will build up behind the emulsion film and cause blistering and loss of adhesion.

7.11 Why do some paints require regular stirring?

To prevent the solid particles of the paint settling out to the bottom of the tin while in use.

Settling-out, or sedimentation in paint occurs when the solid particles are not kept in SUSPENSION by the liquid. This happens when the solids are more dense than the liquid as they have a higher SPECIFIC GRAVITY. This also occurs when the particles are large in size.

Cork and oil float on water because they are both less dense than the water (Fig. 7.11A). A steel girder sinks because it is more dense than water.

Fig. 7.11A

When solid particles are dispersed in a liquid, a good suspension is obtained when the densities of the solids and liquids are similar. Under these ideal conditions, the solids would stay in suspension. Unfortunately, the solid particles in paint are much more dense than the liquid and they settle-out. The VISCOSITY of the liquid helps to prevent sedimentation. Increased viscosity helps to support more dense particles of solid. In ordinary paints, viscosity keeps the solids in suspension. Masonry paints, however, contain large, heavy particles of solid which still settle-out through a viscous liquid and need to be stirred at regular intervals while being used (Fig. 7.11B).

Pigments made from lead are always heavy. Red lead has a relative density, or specific gravity,

PIGMENT SETTLED

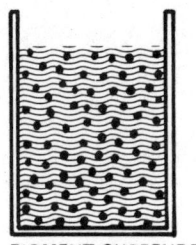
PIGMENT SUSPENDED

Fig. 7.11B

of 8.6 which means that it is 8.6 times heavier than water. When lead pigments are used in paint they tend to sink to the bottom of the tin because they are so dense. Lead paints which have been in store for a long time usually separate. The pigments settle in a hard mass at the bottom of the tin and are difficult to stir back into the medium.

METALS

8.1 What are ferrous metals?	8.6 How is rust removed?
8.2 What are non-ferrous metals?	8.7 Can corrosion be caused by joining together different metals?
8.3 How do metals corrode?	
8.4 What is rust?	8.8 How can rust be prevented?
8.5 What is mill scale?	

8.1 What are ferrous metals?

FERROUS metals contain iron. They include: wrought iron, cast iron and the various sheets, such as mild steel, high tensile steel and stainless steel.

Ferrous metals are made from PIG IRON which is extracted from natural IRON ORE. Various amounts of carbon are added to produce iron and steels. Wrought iron has a low carbon content. Cast iron has a high carbon content (Fig. 8.1).

Steel can be ALLOYED with other metals to make them less likely to rust. For example, chromium and nickel can be added to steel to produce stainless steel. Stainless steel is a steel alloy.

8.2 What are non-ferrous metals?

NON-FERROUS metals have no iron in them. They are all the metallic elements except iron and all alloys *without* iron in them. They include: zinc, copper, tin, lead, aluminium, chromium, nickel, silver, gold, brass and bronze (Fig. 8.2).

Non-ferrous metals are obtained by many methods which are generally more difficult and expensive than the production of iron. Most non-ferrous metals are rather soft in their pure form. Harder metals are obtained by mixing together various metals in their molten state to produce ALLOYS. For example, zinc and copper produce brass.

8.3 How do metals corrode?

Either by an electro chemical process or by a reaction with gases in the air.

Corrosion takes many forms. Metals containing iron or FERROUS METALS corrode by RUSTING. This is a reaction between the metal and oxygen in the air. Water must also be present. For example, rust forms on unprotected iron gutters and pipes (Fig. 8.3A). The rust is porous and exposes the metal underneath the rust to further corrosion until it rusts away completely.

Many of the metals which do not contain iron, the NON-FERROUS metals, corrode because of a reaction between them and gases, such as sulphur dioxide and carbon dioxide, which are present in the air. The tarnishing of silver, and the green

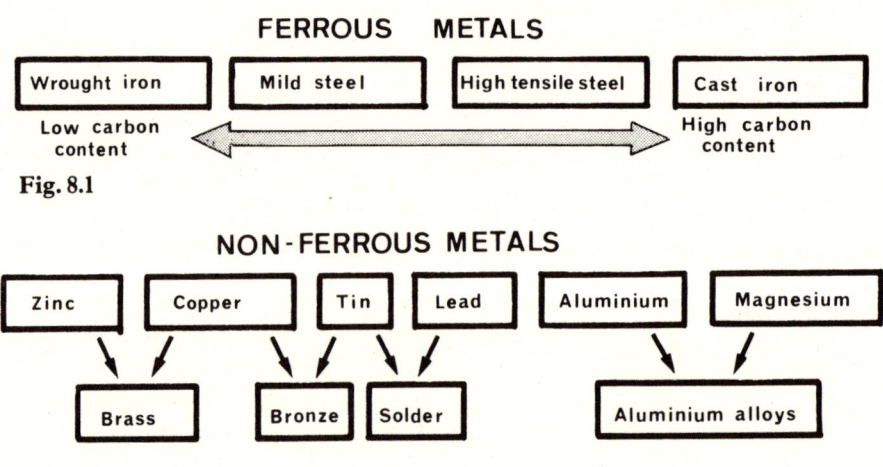

FERROUS METALS

| Wrought iron | Mild steel | High tensile steel | Cast iron |

Low carbon content ⟵⟶ High carbon content

Fig. 8.1

NON-FERROUS METALS

| Zinc | Copper | Tin | Lead | Aluminium | Magnesium |

Brass   Bronze Solder   Aluminium alloys

Fig. 8.2

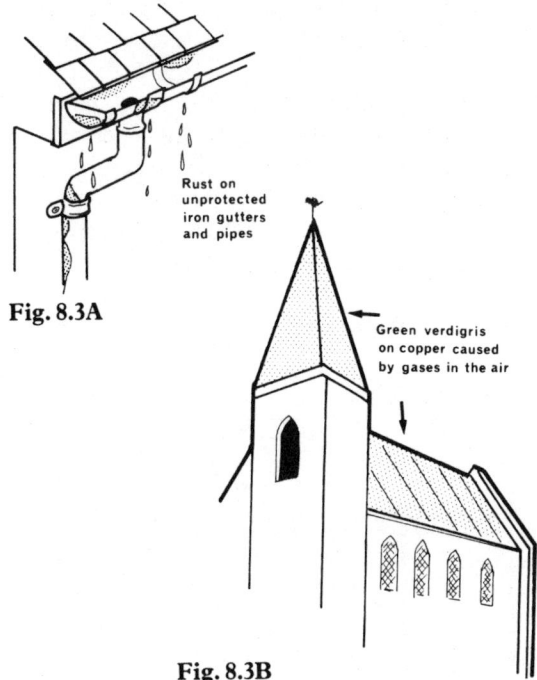

Fig. 8.3A

Green verdigris on copper caused by gases in the air

Fig. 8.3B

to make an ACID solution (Fig. 8.3C). This solution dissolves metals such as zinc. Aluminium can be dissolved by strong ALKALIS like caustic soda.

Polluted wet atmospheres also cause a process called ELECTROLYTIC CORROSION. The metal and the impurities in the air form a SIMPLE CELL. A simple cell usually has two main parts immersed in a chemical. For example, in a flashlight battery, a carbon rod is fixed in a jelly of an ammonium compound which in turn is in a zinc case (Fig. 8.3D, E). This produces an electric current which flows between the carbon rod and the zinc case when they are joined by a wire. The zinc case gradually corrodes.

verdigris which forms on copper, are caused by these gases (Fig. 8.3B). The film that forms on the surface does not flake off like rust. Instead it may protect the metals against further corrosion.

Sometimes metals dissolve without a layer of corrosion appearing. This often happens in very polluted atmospheres around factories. The air contains sulphur dioxide which dissolves in rain

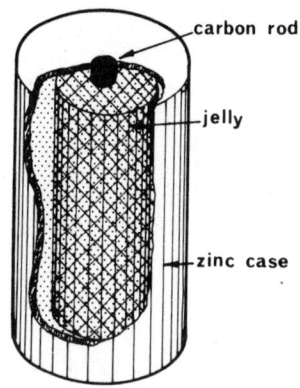

Fig. 8.3D

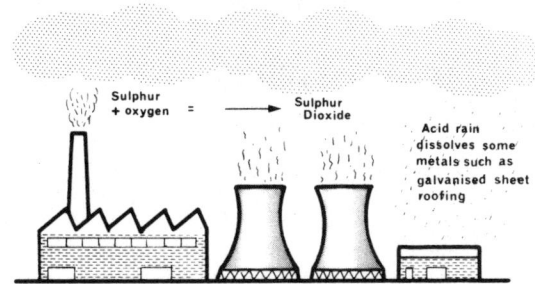

Fig. 8.3C

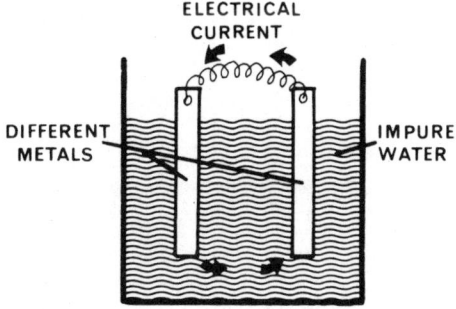

Fig. 8.3E

In the same way, many simple cells are made when impurities and metal are together in a damp atmosphere. An electric current flows and the metal changes into a COMPOUND and so corrodes away.

GALVANIC CORROSION is a similar type of corrosion process. It takes place when different metals are joined together. A simple cell is set up and a current of electricity flows between the metals. The result is that *one* of the metals corrodes. This process is also called SACRIFICIAL CORROSION.

8.4 What is rust?

Rust is a reddish-brown flaky substance which forms on the surface of iron and steel when it is exposed to air and water.

DAMP AIR

Fig. 8.4A

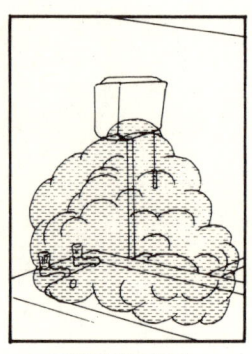

CONDENSATION

Fig. 8.4B

Fig. 8.4C

RAIN

Rust is an OXIDE of iron. This is a CHEMICAL COMPOUND of iron and oxygen. The oxygen comes from the air. In order for iron to go rusty there must be some water present. Damp air, condensation and rain are usually where this water comes from (Figs. 8.4A, B, C).

Rust has no strength and flakes off easily. It is also porous. This means that air and water can always get at fresh iron through the pores of the rust. Eventually all the iron will turn to rust. This is why it is important to prevent rust forming (Fig. 8.4D).

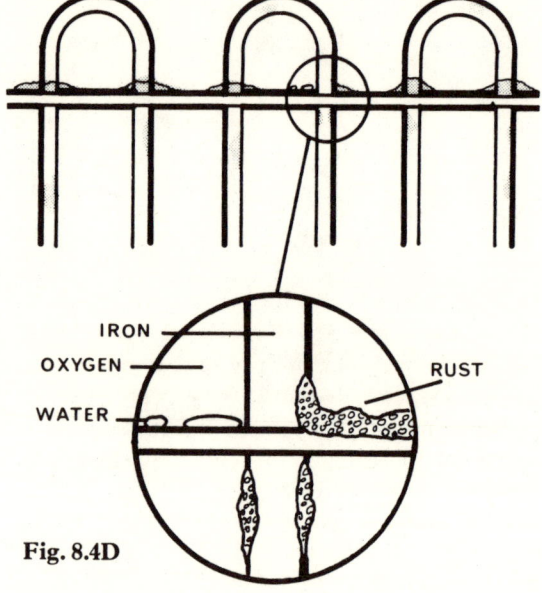

IRON

OXYGEN

WATER

RUST

Fig. 8.4D

8.5 What is mill scale?

A hard, thin crust which forms on the surface of iron and steel while it is being made.

When steel is being made into sheets or beams it needs to be very hot. The temperature may be over 1000°C. During the whole process the steel is in contact with the air and is OXIDIZING. When it comes out of the steel works it is covered with a thin layer of iron oxide or mill scale.

Mill scale has different thicknesses over the surface. When the iron cools down it gets smaller, or contracts, and the thicker parts of the scale, not being very flexible, crack. This is made worse as the steel is knocked about as it is moved to the sites, and left out in all types of weather. Where the mill scale cracks rusting occurs very quickly. Once broken the mill scale is more likely to flake off. It is for these reasons that mill scale often needs to be removed before it is painted (Fig. 8.5A).

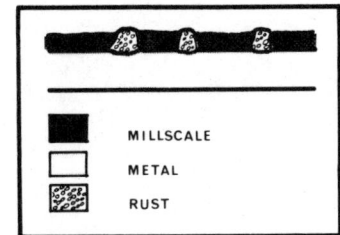

Fig. 8.5A

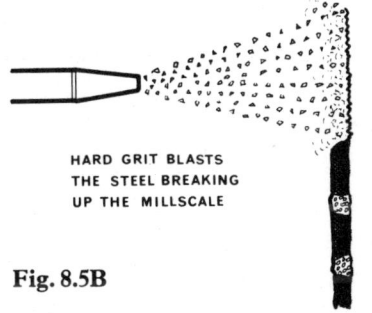

Fig. 8.5B

If mill scaled steel is left out in the weather for a period of between six months and five years, depending on how polluted the atmosphere is, the scale will flake off. When weathering is not possible it can be removed by acid pickling. This requires the metal to be immersed in strong acids which dissolve the softer part of the mill scale and cause all of it to flake off. If the steel is in situ

gritblasting is the only practical method. The hard, sharp grit bombards the surface and breaks up the mill scale, so that it can be brushed off (Fig. 8.5B).

8.6 How is rust removed?

Rust can be removed by dry, wet or hot processes.

The dry process scrapes, brushes or blasts the surface, breaking the rust away from the steel. Abrasive cleaning leaves the metal cleaner than chipping hammers or wire brushes (Fig. 8.6A).

The wet process uses strong acids to dissolve the rust (Fig. 8.6B). The metal is dipped into sulphuric acid, then washed. Finally it is treated

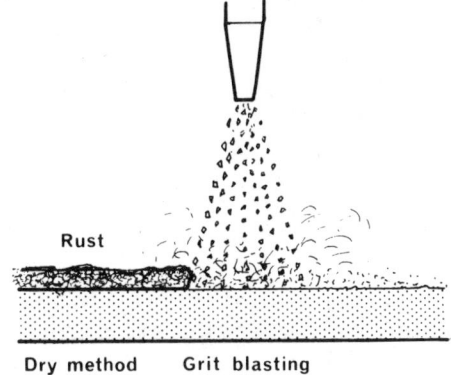

Dry method   Grit blasting

Fig. 8.6A

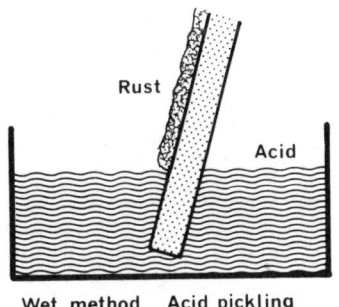

Wet method   Acid pickling

Fig. 8.6B

Hot method   Flame cleaning
Fig. 8.6C

with phosphoric acid. This is called PHOSPHATING. Phosphating leaves the steel surface resistant to rust for a short period before coating.

The hot process uses a very hot flame (Fig. 8.6C). It removes all traces of moisture from the rust. This is called DEHYDRATION. At the same time the heat expands the rust and the steel at different speeds, which causes the rust to break away from the steel. This is known as DIFFERENTIAL EXPANSION. Flame cleaners using oxy-acetylene gas burners are the best forms of heat treatment.

8.7 Can corrosion be caused by joining together different metals?

Yes. In some circumstances a chemical process is started which may result in one of the metals corroding.

When different metals are joined together in the presence of impure water, a SIMPLE ELECTRIC CELL is made. A small current of electricity will flow from one metal to the other. This causes a chemical change on one of the metals and makes it corrode. This type of corrosion is called GALVANIC CORROSION. It could occur, for example, if an iron water tank was connected up with copper pipes (Fig. 8.7A). Galvanic corrosion would affect the

iron tank by making it rust. This is prevented by insulating the joint between the iron and copper with a plastic or rubber washer.

The GALVANIC SERIES shown in table 6 lists metals from ACTIVE at the top to less active or NOBLE at the bottom. When any two metals are joined together the one that is *higher* on the list will corrode. For example, if iron and zinc were joined, the zinc would corrode. This is the principle of GALVANIZING and SHARARDIZING. The zinc slowly corrodes instead of the iron. The zinc gives itself up to protect the iron and this is known as SACRIFICIAL PROTECTION.

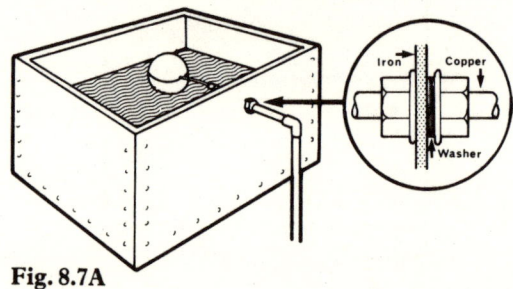

Fig. 8.7A

The further apart metals are on the Galvanic Series the greater is the reaction between them. For example, if aluminium, near the top of the table, is put with silver, which is near the bottom of the table, the aluminium would corrode very quickly. But if aluminium is put with zinc it will corrode much more slowly, because zinc and aluminium are closer together in the series.

The metal which corrodes and offers protection to the other metal is called an ANODE. The protected metal is a CATHODE.

On older buildings iron railings were often bedded into a lead lined hole in the stone plinth. The result was a very fast rusting of the iron at the point where it touched the lead (Fig. 8.7B).

Boats which have a brass screw or propeller can cause steel in the hull or propeller shaft to corrode rapidly. This can be prevented by fixing pieces of

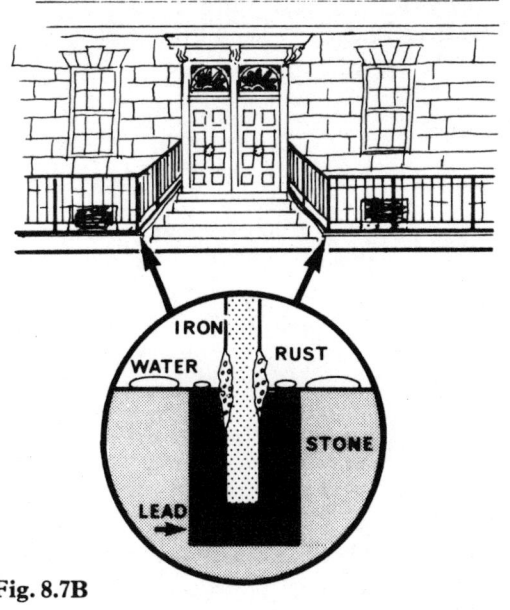

Fig. 8.7B

Table 6 **Galvanic series or electro-chemical series**

Magnesium	
Aluminium	the unstable or active metals
Chromium	
Zinc	
Iron	
Tin	
Lead	
Brass	
Copper	
Silver	the stable or noble metals
Gold	

8.8   How can rust be prevented?

By combining the iron or steel with a non-ferrous metal or by coating it.

There are three ways of preventing iron or steel from rusting.

One is to use a steel ALLOY. This is steel which, during manufacture, is combined with metals which do not rust, such as chromium and nickel. Stainless steel is produced by this process. It is strong, does not rust, but is expensive.

A less expensive way is to use a SACRIFICIAL COATING (Fig. 8.8A). If iron is coated with a metal which is above it in the GALVANIC SERIES, the other metal will corrode first and protect the iron underneath (Fig. 8.8B). Metal coatings include: zinc, which is known as GALVANIZING or SHERARD-IZING; aluminium, which is ANODIZING; chromium which is often called ELECTRO-PLATING.

Once ordinary iron or steel has been used in a building it can be protected from rusting only by coating. By keeping air and water away from the surface of the iron it will not rust. This can be done by painting. Special paint systems are

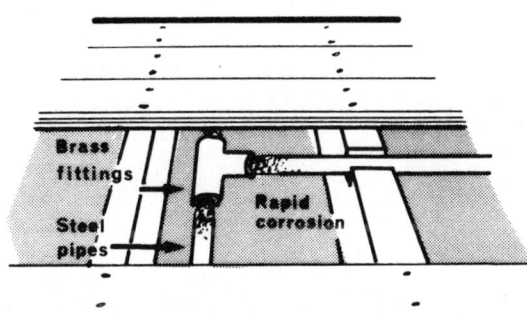

Fig. 8.7C

a sacrificial metal near the screw and these give themselves up instead of the iron rusting. The metal is often zinc and it has to be replaced at regular intervals.

Steel central heating pipes fitted with brass or copper connectors rapidly corrode near the joints (Fig. 8.7C).

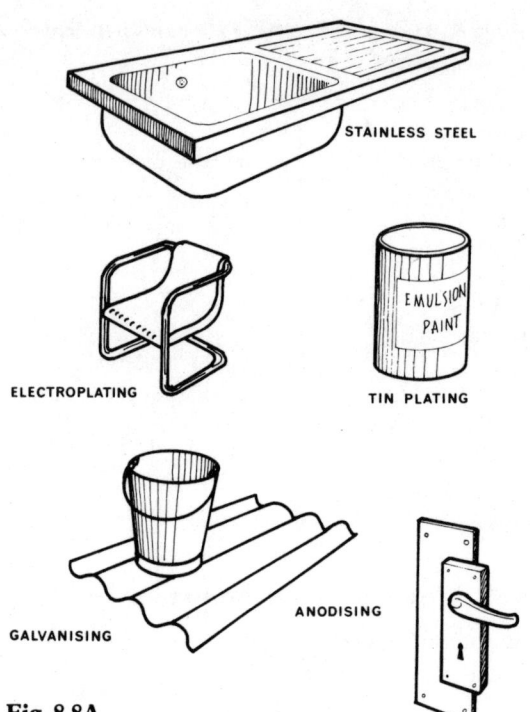

STAINLESS STEEL

ELECTROPLATING

TIN PLATING

EMULSION PAINT

GALVANISING

ANODISING

Fig. 8.8A

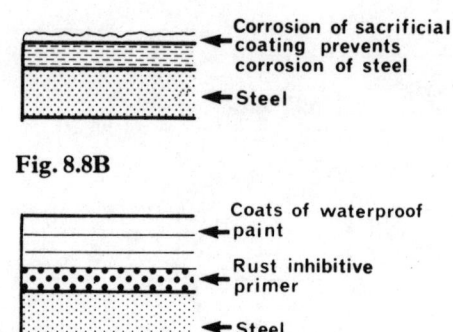

Corrosion of sacrificial coating prevents corrosion of steel

Steel

Fig. 8.8B

Coats of waterproof paint

Rust inhibitive primer

Steel

Fig. 8.8C

necessary. The first coat, or primer, must be made with pigments which are RUST INHIBITIVE. This layer is followed by several coats of tough, waterproof paint (Fig. 8.8C).

Because rust flakes off very easily, it must be removed before any protective treatment is carried out.

PLASTERS AND CEMENTS

9.1 How do plasters and cements set?

9.2 How can the set of plasters be slowed or speeded up?

9.3 Do all cements set and harden at the same rate?

9.4 Why do plasters and cements get hot while setting?

9.5 Why do plasters and cements sometimes set in the bag?

9.6 How does the atmosphere affect the setting of concrete, plaster and mortar?

9.1 How do plasters and cements set?

When water is added to dry plaster and cement, crystals form which gradually join up with each other until it becomes one solid mass.

The raw materials which are used to make gypsum plasters and portland cements are CRYSTALLINE. This means that they are made up of CRYSTALS which are hard, regular shaped particles which fit together.

During the making of plasters and cements heat is used to drive off water from the crystals. This water is known as WATER OF CRYSTALLIZATION. When the dry powder is mixed with water before use, some of the water replaces the missing water of crystallization and new crystals start to form. The new crystals continue to grow and interlock with each other forming a solid or SET material (Fig. 9.1A). This process of crystallization is known as HYDRATION.

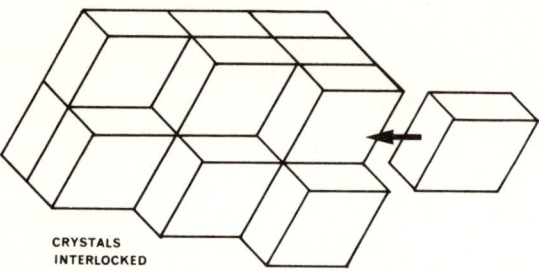

CRYSTALS
INTERLOCKED

Fig. 9.1A

Plasters and cements are HYDRAULIC materials and will set under water. This is why they must not be washed down sinks, because the material will set and block the sink trap (Fig. 9.1B).

The strength of plasters and cements depends on the number of crystals that grow and lock together. They will continue to grow while the material is wet and until all the original water and crystals have been replaced. If the water

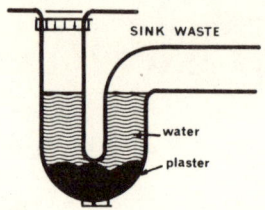

SINK WASTE

water

plaster

Fig. 9.1B

evaporates quickly the hardening may stop before the material has reached its full strength. Therefore, if plasters are dried off quickly they can be very chalky. Cements are usually kept wet for two to three days after being applied to make sure that hydration is complete. After this time the materials are said to be CURED.

9.2 How can the set of plasters be slowed or speeded-up?

By the addition of special RETARDERS which slow the set, or ACCELERATORS which speed-up the set. Uncontrolled changes in the setting time sometimes take place when mixing and pouring conditions are not correct.

When plaster HYDRATES crystals grow and interlock within the material until it sets hard. Retarders, such as glue solution, slow down crystallization by acting as a temporary barrier between crystals, to slow down the rate that they interlock. Plaster of Paris is often retarded with glue solution. Without a retarder it would set in a few minutes, making it difficult to use. Lime is sometimes mixed with plasters to slow the crystallization.

Accelerators work by introducing tiny particles into the mixed material which speed up the crystalline growth. Once crystals are formed they continue to grow and interlock until the material sets hard. Some plasters are made in such a way that without an accelerator they would not set at all.

If freshly mixed plaster comes into contact with old set material, or is mixed with dirty water, small particles are introduced into the mix and the set will be accelerated. A plaster which sets too quickly is difficult to work, and loses strength, because insufficient time is allowed for the crystals to properly interlock. On the other hand, slow setting plasters set hard because of the extra time that allows the crystals to interlock correctly. For this reason it is most important that mixing tools and buckets are cleaned between mixes and only clean water is used.

9.3 Do all cements set and harden at the same rate?

No. Provided the mixing and pouring conditions are correct, all cements will become stiff and unworkable in about forty-five minutes. This is the INITIAL SET. They then HARDEN, which may take some days or years, depending on the type of cement.

The initial set is caused by evaporation of water from the mix and partial HYDRATION. Evaporation of water is fast on hot, dry days and the set will be speeded-up. During cold, damp conditions evaporation is slower, and the set will be slow or retarded.

Partial hydration of the cement and other ingredients in it also stiffens the mix and makes it set.

Special RETARDERS based on sugars and starches can be added to slow-down the set. These work by making a temporary barrier between the forming crystals, which slows down the speed at which they interlock. Freezing conditions totally prevent hydration because ice forms within the wet material and prevents the new crystals locking together (Fig. 9.3).

After the initial set, the cement slowly HARDENS.

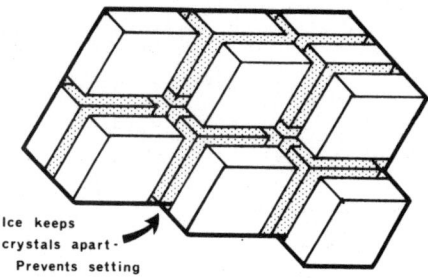

Ice keeps crystals apart - Prevents setting

Fig. 9.3

This is a complicated process of crystals forming and interlocking very closely to form a hard material. When special ACCELERATORS such as calcium chloride are added the hardening process is speeded up. The accelerators cause crystals to form quickly around them. This accelerated hardening produces more heat also and the cement becomes hard in a shorter period.

Contact with old set material or mixing with dirty water may also speed up the set of cement. This is an uncontrolled acceleration which can make the material difficult to use. Clean tools, containers and water must be used for each mix to prevent early hardening.

9.4 Why do plasters and cements get hot while setting?

Because heat is given off when the crystals are formed.

Some chemical changes take place slowly, and the amount of heat released is so small that it is not noticed. Other changes, such as the hydration of plaster and cement, are fast and so much heat is released that the material actually feels hot.

When plasters and cements are made, water is driven-off by heating. But some of the heat energy is *stored* in the dry powder (Fig. 9.4A). When water is added to the powder the mixture begins to SET, and the chemical change that takes

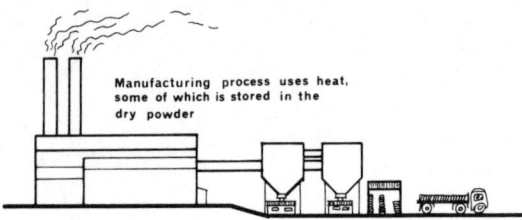

Fig. 9.4A

place releases the *stored* energy in the form of heat. This released heat is called the HEAT OF HYDRATION (Fig. 9.4B).

Processes such as hydration, which release HEAT in this manner are called EXOTHERMIC PROCESSES (Fig. 9.4C). 'EXO' means going out, as in EXIT, and 'THERMIC' refers to heat, as in THERMOMETER.

The hardening of concrete can be slowed by cold weather. If extra cement is used more heat is given off as the concrete hardens which helps to protect it from freezing. Accelerators also may be added which produce more heat as well as increase the speed of hardening and make concreting safer in cold conditions.

Fig. 9.4B

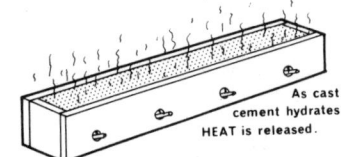

Fig. 9.4C

9.5 Why do plasters and cements sometimes set in the bag?

Because the plasters and cements absorb moisture from the air and start to set.

When water is mixed with plasters or cements they begin to set by a process called HYDRATION. Air contains water in the form of MOISTURE VAPOUR. When a bag of plaster or cement is opened, especially in a damp, unventilated area, it will start to HYDRATE by absorbing some of the water from the air. When this happens the plaster is said to have AIR SET. If cement and plaster is stored for too long, water will penetrate the unopened bag and its contents will air set.

There may be no signs that a bag of plaster or cement has started to air set. They may look in good condition. Yet, when used plaster and cement from old bags may set quicker than material from new bags. This makes them more difficult to use. It happens because hydration has already started. Mixing old and new plasters or cements together may also cause fast setting, so that they are difficult to use.

9.6 How does the atmosphere affect the setting of concrete, plaster and mortar?

Warm, dry atmospheres are ideal for the setting and hardening of plasters and materials containing cement. Heat, cold and excess water can affect the hardening process and possibly damage these materials.

On a hot summer's day, over 25°C, or if heaters are used, plasters can dry too quickly. The hardening, or HYDRATION, process will not have finished before the water evaporates. In this case more moisture will be absorbed later, when the material seems hard, to complete hydration and the surface of the material will expand. This may cause the material to flake off or become powdery.

In very cold condition the water in cements, mortar and plaster can freeze. This will slow down hydration. If this happens before the materials set and harden they may lose some of their strength. They may also flake off or powder.

Water is mixed with plaster and cement for two reasons. One is to make it into a plastic state so that it can be applied to a surface or be used to bond bricks together. The second reason is to make the materials harden by hydration. Any water that is not required for the hydration process will evaporate. If, during hydration, the materials absorb further water from very wet air or rain, they have to get rid of even more moisture before they harden. The evaporation of this excess moisture often leaves many more tiny holes or pores in the plaster and cement than they would normally have. So they dry as very porous materials and lose a lot of their strength.

ACIDS, ALKALIS AND ULTRAVIOLET

10.1 Are building materials acids or alkalis?

10.2 In what ways can acids damage building materials?

10.3 In what ways can alkalis damage building materials?

10.4 Is air an acid or alkali?

10.5 How can it be found out if a building material is acidic or alkaline?

10.6 What effect do ultraviolet rays have on paint films?

10.1 Are building materials acids or alkalis?

Many building materials are alkaline. A few are acidic.

All building materials contain chemicals which may be of an ACIDIC or an ALKALINE nature (Fig. 10.1). When dry this does not matter. But when water is present this acidity or alkalinity may cause a chemical reaction. Alkaline materials are those which contain cement or lime, such as concrete, cement rendering or asbestos cement. Some types of fire retardant solutions are alkaline also. Absorbent surfaces which are in contact with an alkaline surface may absorb alkaline salts in wet conditions. In this way materials which are not normally chemically active may become so. For example, gypsum plaster is not alkaline but when applied over a cement/sand backing it may become alkaline.

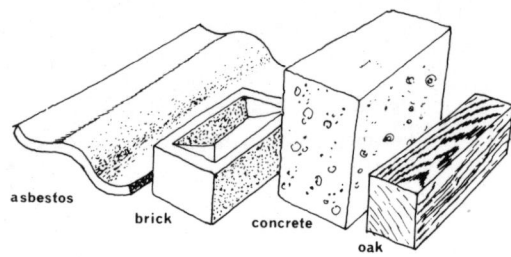

BUILDING MATERIALS CONTAINING DESTRUCTIVE CHEMICALS

Fig. 10.1

There are few building materials which are acidic. Some timbers, such as oak, chestnut and western red cedar are acid.

Sulphate salts from the clay used in bricklaying may produce bricks of an acid nature in damp conditions.

10.2 In what ways can acids damage building materials?

They can cause corrosion and cryptoefflorescence.

If metals, such as iron or steel, come in contact with air or other materials which are acidic a reaction takes place and chemical compounds are produced in the form of powdery deposits on the surface of the metal. This is CORROSION. Steel nails and screws are affected in this way when used in brickwork or acidic timbers such as oak, chestnut or red cedar (Fig. 10.2). If copper and iron are joined together in acidic building materials, a SIMPLE CELL is made and corrosion takes place much faster.

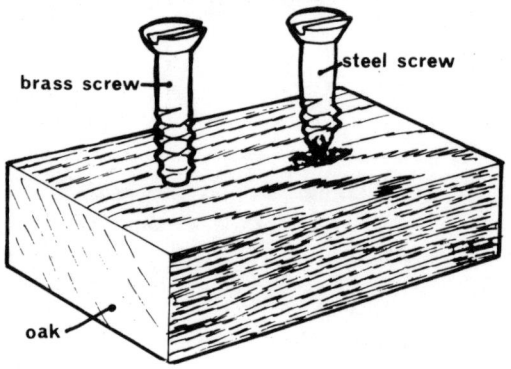

Fig. 10.2

Bricks may be acidic by nature or by absorption of dilute sulphuric acid by rain penetration. This happens more particularly on chimney stacks where acid pollution is high. The mortar is usually alkaline because it contains either lime or cement so a chemical action takes place and crystalline SALTS are produced. The salts expand as they crystallize and create pressures within the pores of the brickwork. The result is that the surface of the brick may flake or the mortar may expand. This is CRYPTO-EFFLORESCENCE.

Cryptoefflorescence is a serious problem. It can cause walls to lean, and ridge tiles and coping stones to loosen.

10.3   In what ways can alkalis damage building materials?

They can destroy some paints, seriously affect colours and reduce adhesion of pastes.

Many oil paints contain linseed oil. This drying oil contains FATTY ACIDS which react chemically to produce SOAPS when applied to an alkaline surface such as cement rendering or concrete. This reaction is SAPONIFICATION and turns the paint into a soft and sticky mass.

Alkalis in surfaces can affect the colour of some pigments. Those pigments which contain lead, such as lead chromes and chrome green are BLEACHED by alkalis. Prussian blue also loses its colours when attacked by alkali.

Alkaline surfaces may react with starch wallpaper pastes which are acidic. If this happens crystalline salts are formed beneath the paper and will push it off the wall.

10.4   Is air an acid or an alkali?

Air can be acidic in certain circumstances.

Air consists of 78% nitrogen, 21% oxygen and 1% of other gases. It also takes up other chemicals from the burning of fuel and manufacturing processes. When coal is burnt sulphur dioxide gas is given off. When this gas dissolves in water vapour in the air dilute sulphuric acid is formed (Fig. 10.4). Paint films break down quicker when attacked by this acid. Iron and steel corrodes because of this dilute acid in the air. Carbon dioxide in the air is also dissolved by rain to form a weak acid solution. This carbonic acid attacks limestone. Many food processing factories use acetic acid for pickling and the manufacture of vinegar and sauces. Around these buildings the air is very acidic. For this reason these fumes must be removed otherwise corrosion of building materials near the factory will be severe.

Fig. 10.4

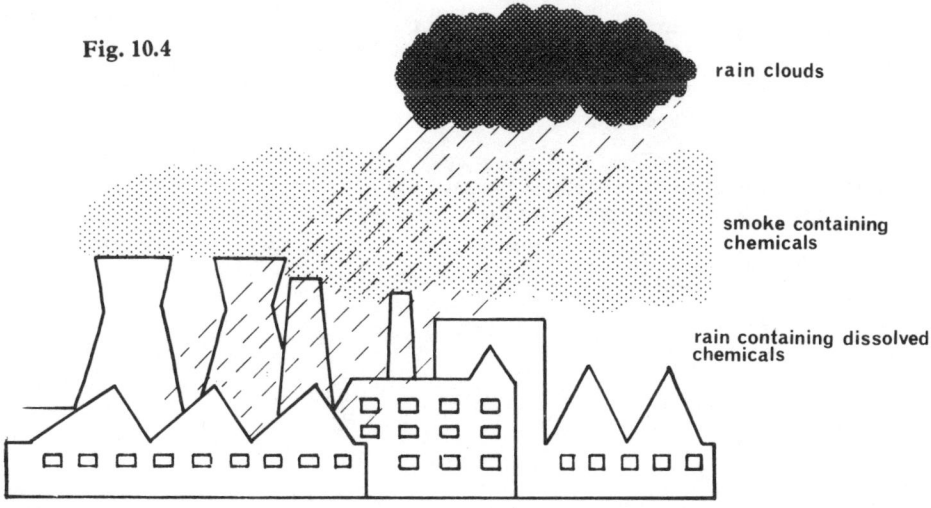

rain clouds

smoke containing chemicals

rain containing dissolved chemicals

10.5 How can it be found out if a building material is acidic or alkaline?

By the use of special indicators in the form of liquids or small paper strips.

There are two main types of indicators.

LITMUS is a vegetable dye which changes colour when in contact with an ACID or an ALKALI. When put in acid the litmus turns red. When put in an alkali it turns blue. The most common way litmus is used is in the form of thin paper strips which have been treated with the dye. These are litmus papers (Fig. 10.5A).

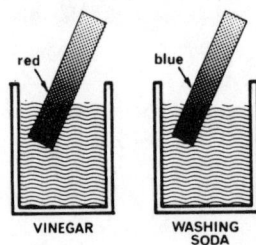

Fig. 10.5A

UNIVERSAL INDICATORS are paper strips which show the degree of acidity or alkalinity of a material also by changing their colour. The colours range from red through yellow, green and mauve, to dark blue, and can be compared with a reference chart.

To use indicators it is first necessary to damp the surface with distilled water to make the chemicals active. Tap water is sometimes slightly acid so should not be used. The indicator paper is pressed onto the damp surface. The universal indicator papers which at first are yellow will change colour according to the nature of the material. The colour indicates whether the material is acid (red), or alkaline (dark blue), or NEUTRAL (green), which is a state between the two. It also tells how strong an acid or alkali it is. This is called the pH VALUE (Fig. 10.5B).

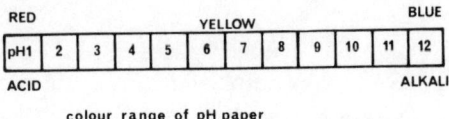

colour range of pH paper

Fig. 10.5B

10.6 What effect do ultraviolet rays have on paint films?

They cause them to break down.

Some of the invisible energy rays of the sun are called ULTRAVIOLET (u.v.) rays. Certain materials, when exposed to ultraviolet rays for a long time, break down. For example, the resins used to make paint films eventually become brittle and lose their gloss after a few years of exposure, mainly because of the action of u.v. Varnish films tend to break down more quickly than gloss paints because they do not have the protection which pigments give paints by reflecting away some of the harmful effects of ultraviolet light.

Paint films on the south facing sides of buildings tend to break down before those on the north facing side. This is because they are exposed to more u.v. rays.

FUNGI

11.1 What causes wood to rot?

11.2 What are fungi?

11.3 What is mould growth?

11.4 How do moulds feed and travel?

11.5 How can wood rot be treated?

11.6 How can mould growth be treated?

11.7 What plants grow on external walls and roofs?

11.8 How can surfaces covered by algae, lichen and moss be treated?

11.9 Why does timber require a protective coating?

11.1 What causes wood to rot?

A FUNGUS which feeds on the cellulose in timber. The growth continues until the timber loses all its strength.

The two main types of fungus which destroy timber are dry rot and wet rot.

When timber is cut from a tree it will contain a large amount of moisture. Most of this must be removed before it can be used in a building. The ideal condition is for timber to have a moisture content of between 12 and 18% of its total weight. This is known as seasoned timber (Fig. 11.1A). If timber can be kept in this state by letting the air get at it and protecting it from moisture it will not be attacked by rot. When it does become wet, with a MOISTURE CONTENT of over 20%, the fungus spores will be attracted to it.

Dry rot is so named because after the fungus has taken all it needs from the timber it leaves it dry and crumbly like a piece of brittle sponge. The dry rot fungus may attack wood which has a moisture content of above 20% (Fig. 11.1B, C).

Wet rot will grow most readily on very wet timber with a moisture content of over 50%. But the wet rot fungus can grow on timbers which have a moisture content as low as 25%.

The main causes of dry and wet rot are as follows (Fig. 11.1D).

(a) Rising damp in walls, particularly when ventilation below the floor is poor. Floor joists and roof rafters are affected most by rising damp.
(b) Where leaks have occurred in houses. This is most likely in the roof, bathroom, or near faulty gutters and down pipes, particularly where the paint has broken down giving timber no protection. All types of joinery may be affected.
(c) The use of unseasoned timber in new houses.
(d) Timber in new houses which has been stacked in the open before fixing and has become very wet.

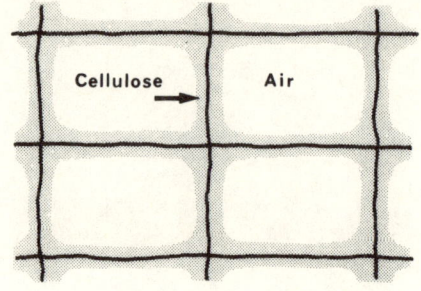

SOUND TIMBER CELLS

Fig. 11.1A

Fungus feeding on wood cells

Fig. 11.1B

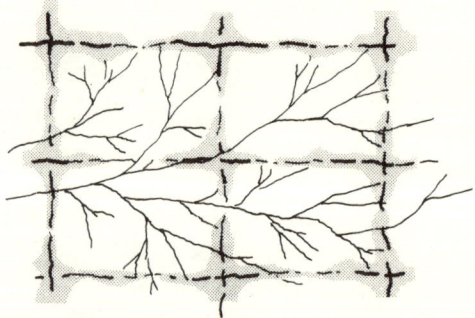

Timber becomes dry and weak

Fig. 11.1C

Fig. 11.1D

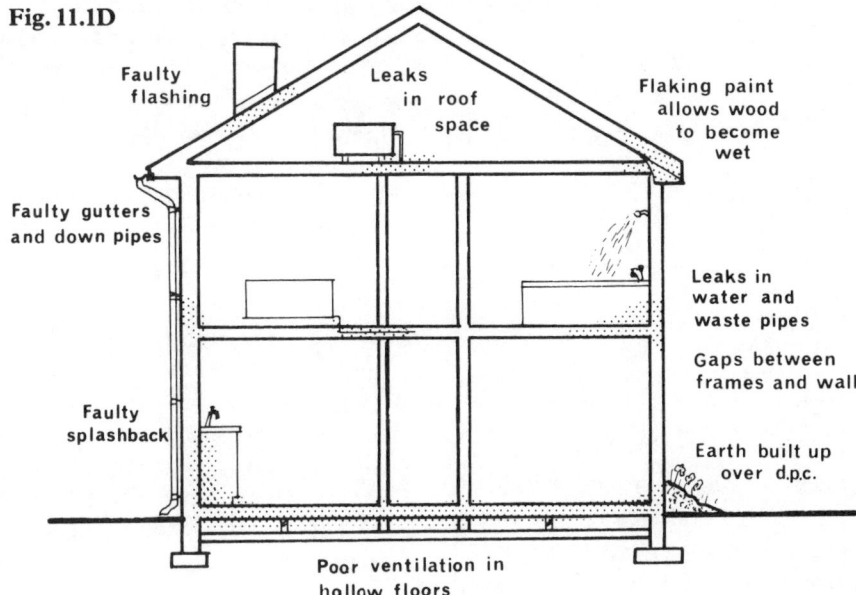

Faulty flashing

Leaks in roof space

Flaking paint allows wood to become wet

Faulty gutters and down pipes

Leaks in water and waste pipes

Gaps between frames and wall

Faulty splashback

Earth built up over d.p.c.

Poor ventilation in hollow floors

11.2 What are fungi?

Fungi are a low form of plant life which has neither stem, root or leaves.

There are thousands of different types of fungi. One form is the mould that appears on walls and ceilings of buildings. The worst kinds are the WET and DRY ROT fungi which feed on, and destroy, timber. Mushrooms are a type of fungi also.

Unlike other plants, fungi cannot produce their own food. They need to feed on ORGANIC matter. They also require moisture, and they reproduce faster in dark, warm places.

11.3 What is mould growth?

A type of fungi.

There are many types of moulds which grow mainly on damp internal building surfaces. It is sometimes found on external surfaces. When they start to develop they look like black, green, and, occasionally, purple, spots of dust. As the mould grows the spots join up until the surface is covered. Viewed through a magnifying glass the fine threads spreading out from the mould can be seen.

Like all fungi, moulds must have dampness and food to live (Fig. 11.3A). The dampness can be caused by rain penetrating through walls or roofs, or rising up through the walls from the ground, if the damp proof course is broken or bridged. The most common cause is condensation. Moulds can start to grow in new buildings before they dry out. Moulds do not need light to grow. Food for mould is usually carried to the surface in condensation. It can be dust, or cooking greases, or small bits from food-making processes in factories. Wallpaper starch paste, and some oil paints, supply the food they need also.

101

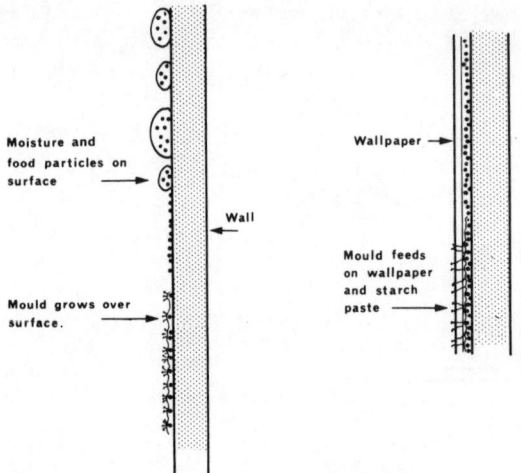

Fig. 11.3A   **Fig. 11.3B**

Mould growth is most commonly found in breweries, dairies, bakeries, factories preparing food and kitchens where the walls may be covered with food particles and the atmosphere is damp and warm. They will be found also in dark places and where ventilation is poor, like in or behind cupboards, behind curtains, in dark corners and behind wallpapers (Fig. 11.3B).

11.4   How do moulds and fungi feed and travel?

Moulds are a type of fungi. Fungi reproduce themselves by producing SPORES which are very tiny seeds. Each fungus plant gives off millions of spores and, as they are so small and light, they can be easily carried in the air or by insects. With so many spores produced it is certain that some of them will find a damp surface with the right food supply on which to feed and grow.

Fungi cannot produce their own food like other plants, but must obtain it ready made. There are thousands of different fungi and they all feed on different foods. They like natural materials, which are called ORGANIC, such as paper, leather, paste and vegetable oils. They will grow on dead and living trees, plants and fruit.

Moulds particularly like surfaces which have a thin film of organic matter on them. These are found on most walls of kitchens or food factories or breweries, carried there as dusts or in condensation.

As well as surfaces which offer the right food, fungal spores need the following conditions.

(a) Air which has about 75% RELATIVE HUMIDITY. They grow extremely fast in tropical countries.
(b) An air temperature of 22–26°C. Fungi develop very quickly in warm conditions. They can survive very cold or freezing temperatures and develop fast when conditions improve.
(c) Most types grow better with little or no light.
(d) Poor flow of air encourages growth.
(e) Rough and soft surfaces are ideal for spores to cling to.

11.5   How can wood rot be treated?

The infected timber must be cut out and burnt, and the surrounding areas must be STERILIZED.

DRY ROT is more difficult to treat than WET ROT. Every trace of affected wood must be removed. Not just the wood that looks rotten but at least 300 mm into what seems to be good timber. This is because the dry rot fungus sends out strands ahead to start attacking new timber. If these are not removed the fungus could continue to grow. All infected timber must be burnt at once to destroy the fungus spores and the strands. Infected areas, including adjoining brickwork and plaster, must be sterilized, either by burning with a gas torch or by applying a special FUNGICIDAL SOLUTION (Fig. 11.5). Fungicidal solutions contain poisons which kill fungus. The remaining timber, and the new timber, must be

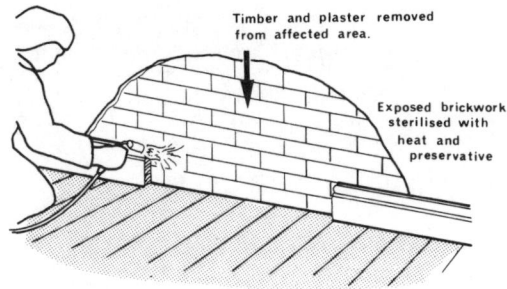

Timber and plaster removed from affected area.

Exposed brickwork sterilised with heat and preservative

Fig. 11.5

treated with wood preservatives. Fungal SPORES will still settle on preserved timber but they will quickly die because the timber can no longer offer them the type of food they need.

When treating WET ROT it is necessary to stop the water getting to the timber. When the timber and areas around it have dried out, the wet rot fungus will die and the rotting will stop. All decayed timber must be cut out and burnt to kill existing fungus. Replacement timber must be treated with a preservative before fixing to prevent any new spores growing should the timber become wet again.

11.6   How is mould growth treated?

By destroying the mould with chemicals or heat.

Mould can be killed by being washed with household bleach or carbolic acid. But when the surface becomes dirty, new SPORES might grow again. STERILIZING SOLUTIONS contain substances which kill the mould. They also leave traces of the TOXIC material on the surface so that new spores will not grow on it. Moulds can also be killed by being burnt with a gas torch.

11.7   What plants grow on external walls and roofs?

Algae, lichens and moss.

These are low forms of plant life, commonly seen on roofs and external walls of buildings, particularly in country areas. All three must have light, a food supply and moisture to survive.

Algae. There are many different types, most of which are found in or near the sea or rivers. The stem, leaf and root is all in one and they do not require soil. A few types are found growing on very damp or wet walls, especially near broken gutters or down pipes.

Lichens are a mixture of fungi and algae. The algae part makes the food while the fine threads of fungus obtain the water, often from surfaces too dry to support algae. They do not require soil and are often found on rocks, trees, natural stone walls and tomb stones, where they grow over, and hide, the carvings and inscriptions (Fig. 11.7).

Lichens, a mixture of fungus and algae do not require soil

Fig. 11.7

Moss. Unlike algae and lichens, moss has a primitive root system and requires a fine layer of soil to supply the moisture and salts they need. They also require high HUMIDITY or damp conditions. Commonly found in woods, forests and mountainous areas, they can also be found growing on roofs and in gutters where they can cause blockages.

Unlike fungi, these plants rarely destroy a surface. Many people think they look nice on surfaces, especially if the building is old. Where there is a contrast between old and new surfaces on repairs and extensions, this growth is sometimes encouraged to blend the two together.

11.8 How should surfaces covered by algae, lichens and moss be treated?

By applying a TOXIC wash to kill the growth.

There are a number of proprietary TOXIC WASHES available, containing special chemicals to kill the growth. Household bleach can also be used. Some growths, particularly algae, may cause disfiguring stains on masonry, brickwork or rendering, which cannot be removed.

When a contaminated surface is to be cement rendered or painted it is not sufficient to just kill the growth. All the dead matter must be removed so that the adhesion of the rendering or paint is not weakened.

11.9 Why does timber require a protective coating?

To protect it against destructive attack from wood-boring insects and wood rot.

Wood-boring insects. WOODWORM is a general term used to describe an attack by various types of beetles (Fig. 11.9A).

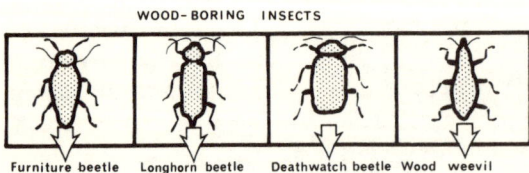

WOOD-BORING INSECTS

Furniture beetle   Longhorn beetle   Deathwatch beetle   Wood weevil

Fig. 11.9A

The adult beetle lays its eggs in cracks and crevices of unprotected wood. The eggs hatch, and small grubs bore into the wood. They tunnel to and fro, feeding on the starch and sugar content of the SAPWOOD (Fig. 11.9B, C). Eventually there may be so many tunnels in the wood that it becomes weak. Structural timbers, like roof rafters and floor joists, become unsafe if seriously infected with woodworm (Fig. 11.9D).

Paint and varnish coatings and poisonous wood preservatives prevent the beetle from laying its eggs in the wood.

Wood rot. The dry and wet rot fungus grows in and destroys wet timber.

Wood is a porous material, full of tiny air

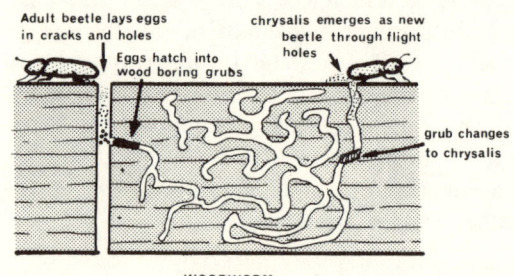

Adult beetle lays eggs in cracks and holes   Eggs hatch into wood boring grubs   chrysalis emerges as new beetle through flight holes   grub changes to chrysalis

WOODWORM

Fig. 11.9B

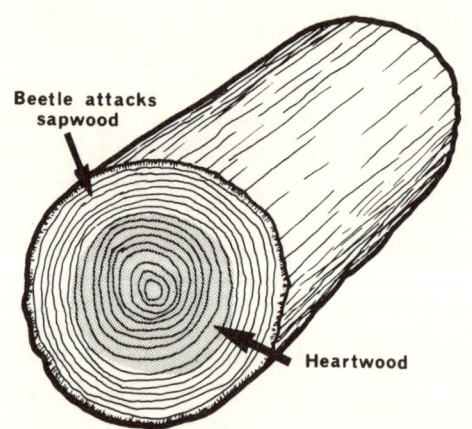

Beetle attacks sapwood   Heartwood

Fig. 11.9C

spaces called *voids*. Many of these voids are linked together. If the surface of wood gets wet, water is absorbed into the voids and passes from one to another by CAPILLARY ATTRACTION. The end grain of wood is very absorbent and water enters easily (Fig. 11.9E). If the moisture content of wood is kept below 20% the spores of wood rot cannot survive.

Wood is given a protective coating of paint or varnish to prevent it absorbing water and fungus cannot live in it. Wood preservatives are FUNGICIDAL. This means that they kill any fungus spores which may settle on the wood.

Structural timbers become unsafe if seriously infected with woodworm

Fig. 11.9D

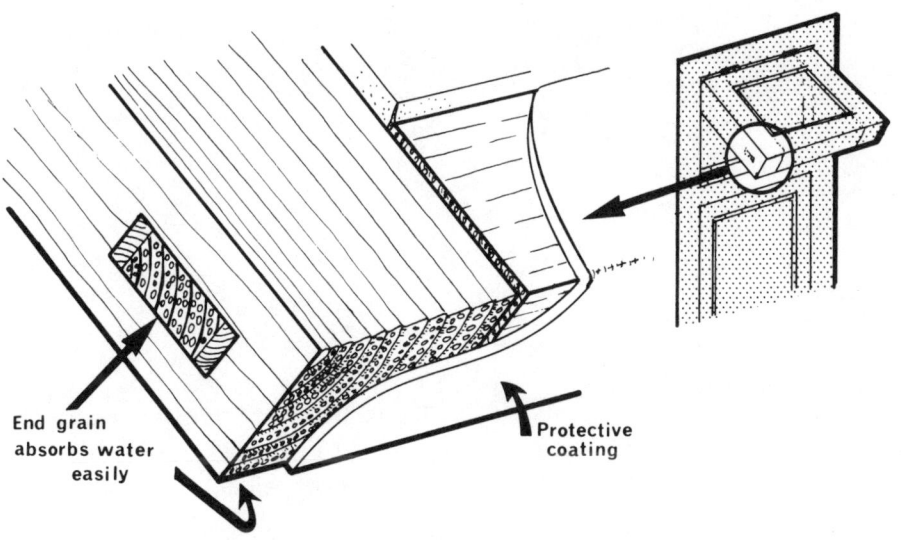

End grain absorbs water easily

Protective coating

Fig. 11.9E

GAS

12.1　What is liquefied gas?

A gas which has been changed into a liquid.

All known gases can be LIQUEFIED provided they are made cold enough and/or put under pressure. Most gases which are available in liquefied form have been made both cold and pressurised. The degree of coldness and the amount of pressure needed to liquefy them varies with each gas (Fig. 12.1A).

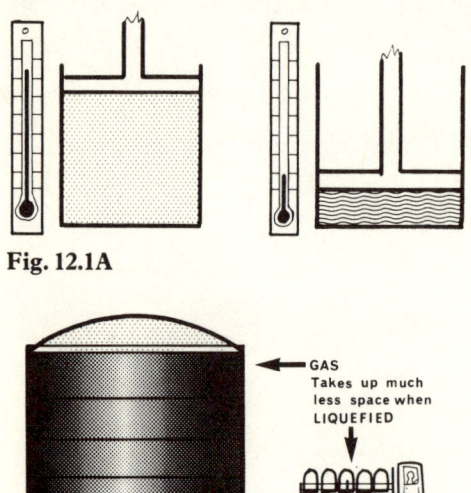

Fig. 12.1A

Fig. 12.1B

In a liquefied state gases take up much less room (Fig. 12.1B). This makes them easier to transport and store. A liquefied gas can expand to about 250 times its volume when it changes back to a gas. Liquefied gases are kept in strong metal containers. When the pressure on the liquefied gas is released it absorbs heat from the container and changes back to a gas. A simple valve on the cylinder controls the quantity of gas released.

The temperature at which the liquids change into a gas is their boiling point. The boiling point of liquefied gas is very much lower than the boiling point of water. If the air temperature is below the liquid's boiling point it will not absorb the heat it needs to turn into a gas, and so cannot be used. Butane has a boiling point of 0°C. So if the air temperature is below freezing point butane will not turn into a gas if the cylinder valve is opened (Fig. 12.1C). The boiling point of propane is —47°C.

Air temperature also affects the pressure at which the gas leaves the cylinder. This is the vapour pressure. Vapour pressure increases with air temperature. On a hot day the gas will come out of the cylinder with more force.

The most common form of liquefied gas is Liquefied Petroleum Gas, LPG. Two kinds of LPG are mostly used, propane and butane. They

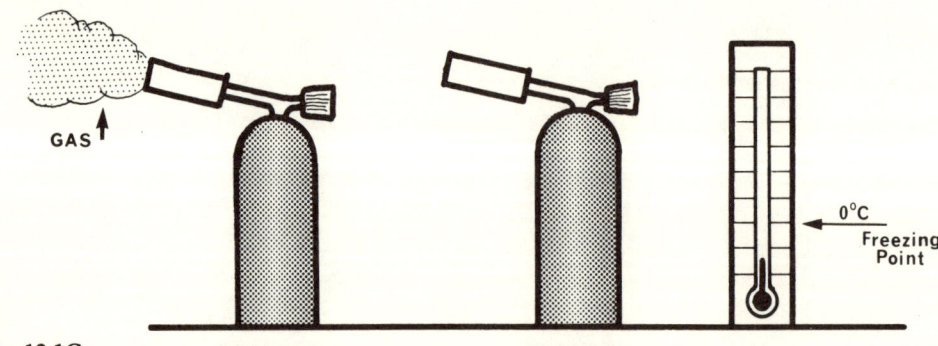

Fig. 12.1C　　PROPANE　　　　BUTANE

are both flammable gases and widely used for operating heating and lighting equipment.

Both propane and butane have a RELATIVE DENSITY greater than air so fall to ground level when released (Fig. 12.1D).

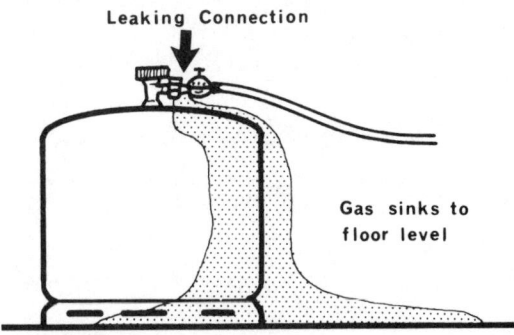

Fig. 12.1D

12.2   How do aerosols work?

By applying pressure on to a liquid contained in a sealed canister to force it through a small nozzle and into the atmosphere. As the liquid leaves the nozzle it expands and is split into minute particles in the form of a fine vapour, or spray, which is called ATOMIZATION (Fig. 12.2).

The canister is made to stand high pressures from inside. It is part filled with the material to be sprayed. The remaining space contains a small amount of a liquefied gas, which is known as the PROPELLANT. The propellant vaporizes at temperatures above freezing, which causes it to expand in the space above the material. This expansion produces a pressure inside the canister

which is higher than the pressure of the atmosphere. This pressure on the surface of the material pushes it through the supply tube and out through the nozzle when the spring valve is pressed down.

When pigmented materials are used, the canister contains a small steel ball which helps to mix the material when shaken.

Aerosols are used for small quantities of paint, lacquer, preservative and some adhesives. They have a wide use in cosmetics in the form of deodorants and hair sprays.

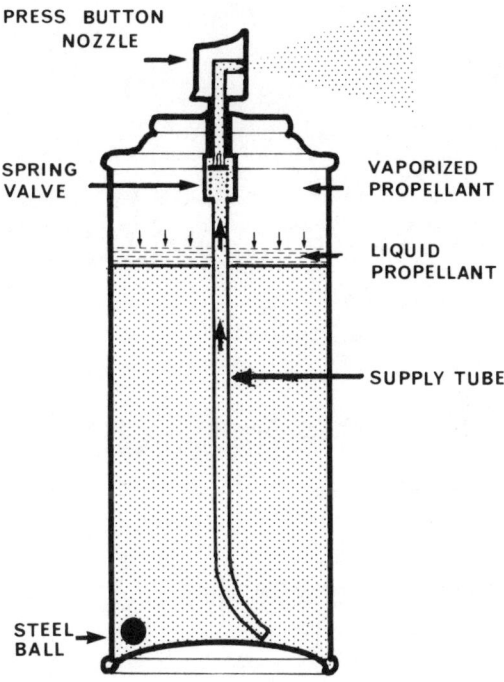

Fig. 12.2

SOUND

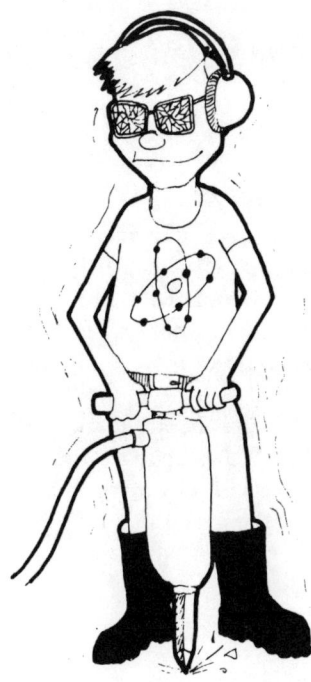

13.1 What is sound?

Sound is a vibration. The sounds we hear are vibrations of air.

When a hammer hits a steel chisel, the chisel vibrates, making the air around it vibrate. These vibrations are the air repeatedly compressing and expanding in *ripples*. As ripples move across a pond when a stone is thrown in, so the *ripples* in the air move until they meet an object, making it vibrate (Fig. 13.1A). In the ear, the eardrum vibrates, followed by various small bones in the ear sending messages along the auditory nerve to the brain. This is experienced as sound (Fig. 13.1B).

Fig. 13.1A

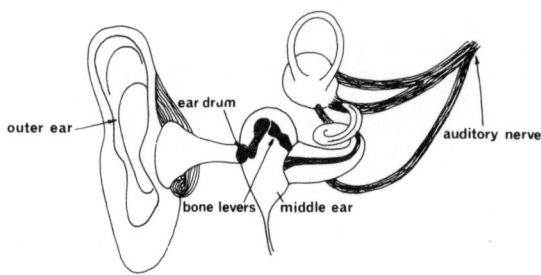

Fig. 13.1B

The more vibrations there are in a certain time, the higher the note we hear. This is called FREQUENCY. The highest note we can hear is about 18 000 vibrations per second. Loud sounds are caused by lots of air vibrating. This is called VOLUME. Some frequencies and high volumes of sound are harmful to our ears. This is why ear protectors are worn when doing noisy jobs (Fig. 13.1C).

Fig. 13.1C

Sounds can travel through liquids and solids as well as through air. Sounds can be heard under the water in a swimming pool. They can travel through brick walls from one room to another. Sounds pass through a window more easily than through a wall.

The speed of sound varies. In air, at sea level, sound travels faster than higher up in the thinner atmosphere. Sound travels faster through water, and faster still through the ground.

13.2 What is echo?

The bouncing of sound waves off hard surfaces.

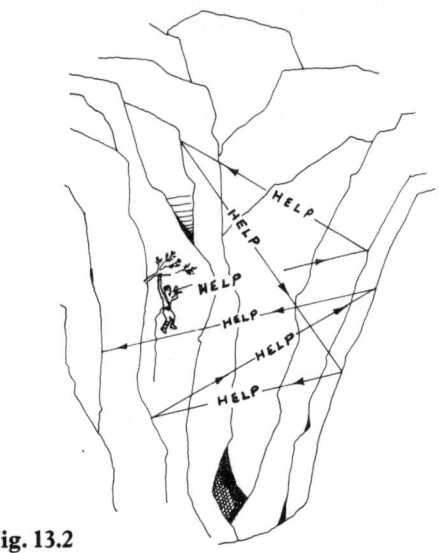

Fig. 13.2

When sound waves travelling through air strike against a hard surface they are REFLECTED back. Just as a torch beam is reflected when shone onto a mirror. The reflected sound is heard again. This is an ECHO.

Sometimes the reflected sound waves are sent off in another direction until they meet another hard surface, when they may be reflected again. This can happen many times, very quickly, and is known as multiple echo or REVERBERATION (Fig. 13.2).

Echoes and reverberations occur in mountain areas, in city streets and in empty rooms.

13.3 What is sound insulation?

The use of certain materials to reduce sound REVERBERATION.

When an echo occurs it may be repeated for two or three seconds. Therefore, it is very difficult to understand talk in a room which has echoes because sounds made a few seconds earlier can be heard, as well as the present sounds.

A room is said to have good ACOUSTICS if echoes are reduced or removed, and sound is directed to where it can be heard most easily.

An empty room echoes. Once furniture, curtains and carpets are put in the echoes seem to stop. The sound is absorbed. This happens because when the sound waves strike fabrics they are bounced back and forth between the fibres of the material until their energy is used up.

Another common acoustic, or sound insulating material, is textured soft boarding. As the material is soft, sound waves are not so readily bounced off the boarding. What little waves are left are bounced between the projections on the board until all their energy is lost (Fig. 13.3A).

Silencers on cars are devices designed to trap the sound from the engine in a series of tiny compartments (Fig. 13.3B).

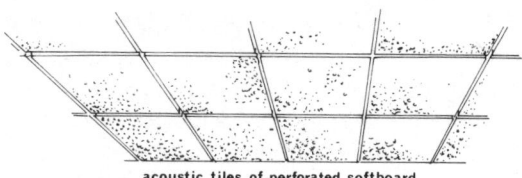
acoustic tiles of perforated softboard

Fig. 13.3A

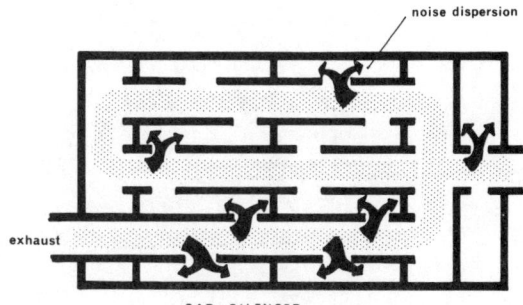
noise dispersion

exhaust

CAR SILENCER

Fig. 13.3B

Concert halls are designed so that sounds produced on the stage are directed to the audience. Whatever sounds reach the walls are absorbed into the crevices of the insulating materials and not reflected as echoes.

Sound is a form of energy. The more energy that is absorbed the less sound is heard. Inside very old buildings it is usually quieter than in some modern buildings. This is due largely to the walls being much thicker and outside noises cannot penetrate them so easily. The material in the wall absorbs all the sound vibrations.

13.4 How is sound measured?

The unit of sound measurement is called a DECIBEL (dB).

It is a scale based on the sound that the ear is receiving. The ear is sensitive over a large range of sound energy. The decibel scale relates to the intensity of these different sounds and arranges them in an order from soft to loud.

Zero on the decibel scale is the faintest sound which an average ear can hear. The sound of rustling leaves is 20dB. The noise level of normal conversation is 40dB. A vacuum cleaner registers 60dB. The sound of a diesel engine registers 80dB and the sound of a pneumatic drill 90dB.

Above 120dB the noise is so loud that it is painful to listen to. A jet engine at full power heard at 30m is rated 140dB (Fig. 13.4). A person standing this close to a jet will need to wear proper ear protection otherwise severe hearing damage may occur.

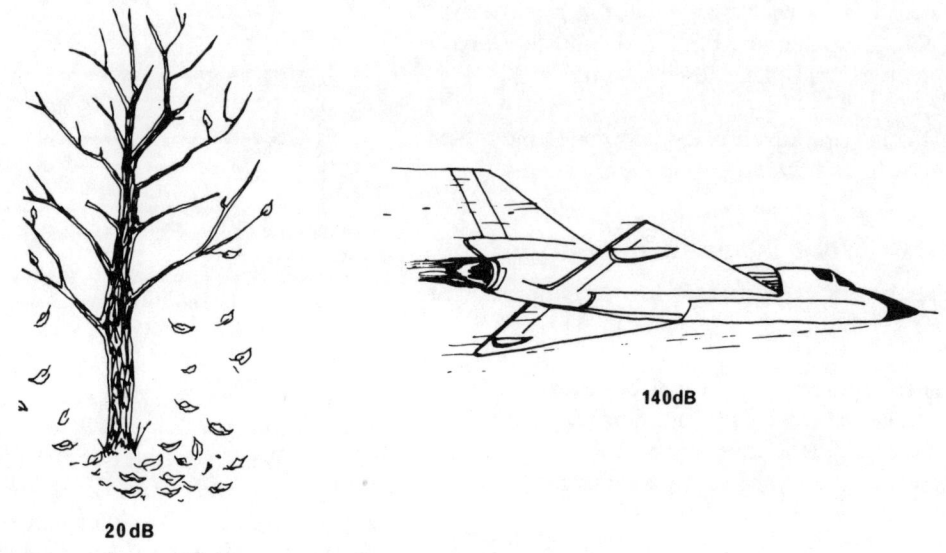

20 dB

140dB

Fig. 13.4

LIGHT AND COLOUR

14.1 What is light?

14.2 What kinds of artificial light are available?

14.3 What are the characteristics of artificial lights?

14.4 What are the effects of artificial light on colour?

14.5 What is used to measure light?

14.6 In what way does the colour of walls affect the brightness of a room?

14.7 How does light pass through some materials but not others?

14.1 What is light?

Light is a form of energy which can be seen by the eye.

The sun gives off many different kinds of energy. Most of the energy is stopped by the atmosphere surrounding the earth, but some comes through. Over millions of years, forms of life have developed parts of their bodies to detect one small part of this energy. The part of our body, and that of many animals, that sees the energy is the eye. The energy that we see is known as visible light.

The sun also gives off radio waves, infra-red radiation which we feel as heat, ultraviolet radiation which tans the skin, and other radiation. The eye is not sensitive to these. We do not see them but they are there.

Just as energy from the sun can be divided into different kinds, so light can be divided into different sorts, or colours. White light is a mixture of colours (Fig. 14.1). Each colour has a different energy.

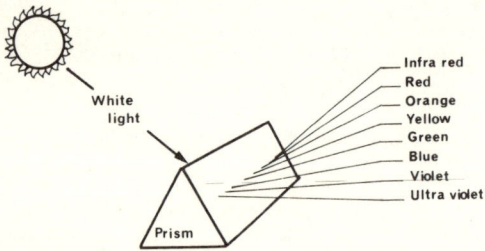

Fig. 14.1

White light contains red, orange, yellow, green, blue, indigo and violet colours. These colours are the SPECTRUM.

Light can be made artificially in many ways. Usually, an object is heated to produce light. In a candle, the burning wax produces carbon which glows with a yellow light when hot. In light bulbs a piece of tungsten wire is heated by an electric current until it glows white hot.

14.2 What kinds of artificial light are available?

There are three kinds in general use. They are tungsten, discharge and fluorescent.

A tungsten bulb is known also as a filament lamp (Fig. 14.2A). Ordinary domestic bulbs are of this type. The lighting filament is made of tungsten wire. The wire glows white hot when sufficient electrical current is passed through it. The filament is protected by a glass bulb. All the air is taken out of the bulb so that the filament does not combine with oxygen and burn out.

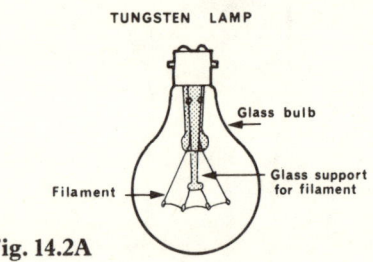

Fig. 14.2A

Discharge lights are glass tubes filled with either sodium or mercury vapour (Fig. 14.2B). Metal plates fitted at each end of the inside of the tube let a stream of electricity flow along the tube. This electricity knocks off other particles of electricity from the outside of the vapour atoms. When these particles join back on a luminous glow occurs.

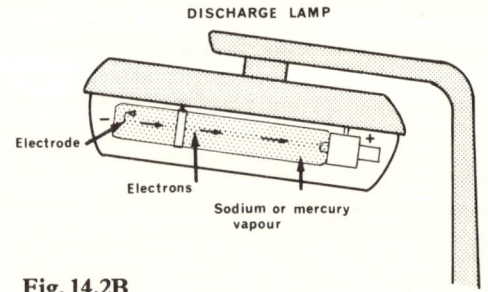

Fig. 14.2B

The type of gas used in the lamps affects the colour of the light. Sodium gas glows with a bright yellow light. Mercury vapour has a bluish white glow.

Discharge lights are used for street lighting, loading bays and floodlighting. They are not suitable for use in homes because of the poor colours they produce.

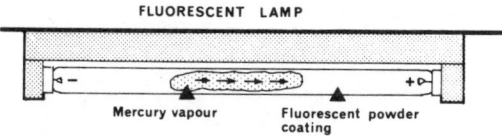

FLUORESCENT LAMP

Mercury vapour   Fluorescent powder coating

Fig. 14.2C

Fluorescent lamps work in a similar way to discharge lamps (Fig. 14.2C). The glass tubes contain mercury vapour. This vapour gives off ULTRAVIOLET light when the flow of electricity collides with it. The inside of the tube is coated with a FLUORESCENT powder which glows when the u.v. light strikes it. Different types of powders are used to coat the inside of the tubes to produce a wide range of lights. Although fluorescent lamps produce light mostly around the blue and violet part of the SPECTRUM, warm pink lights are available also. Fluorescent lamps are widely used in houses, offices, factories and schools.

14.3 What are the characteristics of artificial lights?

Artificial lights differ in their cost and their colour.

Tungsten lamps waste a lot of electricity as about 96% of the energy they use is converted to heat. Fluorescent lamps are more efficient. They give out more light than tungsten lamps using the same amount of electricity. About 12% of the energy used by fluorescent lamps is converted to light. The amount of electricity used is measured in WATTS. The amount of light given off is measured in LUMENS. Table 7 shows the lumen output of lamps for the amount of electricity used.

The colour of artificial light varies according to the way in which the light is made.

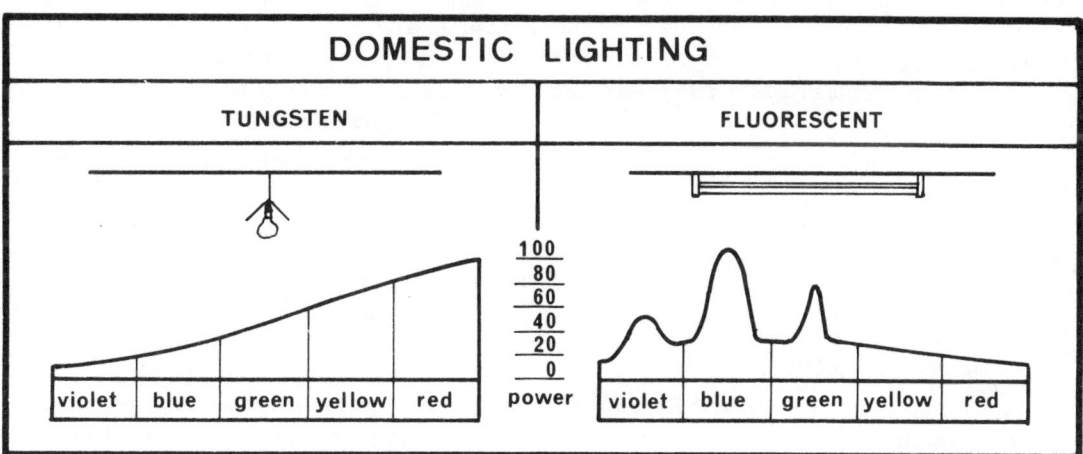

Fig. 14.3

117

Table 7 **Properties and uses of artificial light**

Type	Lumens per watt	Main use	Characteristics
Tungsten filament	5–8	Domestic lighting	Cheap to instal Yellowish light Generates heat
Fluorescent	25–80 depending on colour of tube	Wide use from domestic to special use such as colour matching	Expensive to instal
Sodium discharge	120–175	Street lighting Car parks	Cheap to run. Turns dark colours black, light colours grey Turns yellow orange and reds brown
Mercury fluorescent	35–55	Commercial Industrial	Turns greens slightly yellow and yellows slightly green

White light is made of many colours. The whole range is known as the SPECTRUM. Some lamps give out one or more of the colours in the spectrum more strongly than the rest of the colours. For example, sodium lights are very yellow. Tungsten lights give out more yellow and red than blue.

This range of colours from the lamps and the intensities of the colours is the SPECTRAL RANGE. Each type of artificial light has its own spectral range (Fig. 14.3).

14.4 What are the effects of artificial light on colour?

The colour of a surface will appear different under light from different lamps.

The colour of a surface depends on the type of light which falls on it and the ability of the surface to absorb or reflect the light. A surface which looks green in natural light will look black under red light. A red surface in daylight will look black under green lights.

The nature of artificial light depends on the type of material used to produce the light. They are tungsten, fluorescent powders, mercury vapour or sodium. The light given off by these materials is different in colour. For example, the light from a tungsten lamp is more powerful at the red and yellow end of the spectrum than at the blue and green end. This can affect the colour of a surface under a particular light. A yellow surface absorbs blue light and other colours and reflects yellow only. When seen under a tungsten light a yellow surface looks bright and intense. A blue surface which absorbs yellow and reflects blue loses its brightness and looks dull when seen under tungsten light. The yellow light is almost completely absorbed by the surface. There is very little blue light coming from the tungsten lamp to be reflected.

When a yellow surface is seen under a sodium light only a small amount of light is absorbed and the surface appears bright (Fig. 14.4). Because most main roads are lit by sodium light many emergency vehicles are painted yellow in order to show up well. All other colours look brown or black.

Table 8 shows how colours are affected when seen under different lights.

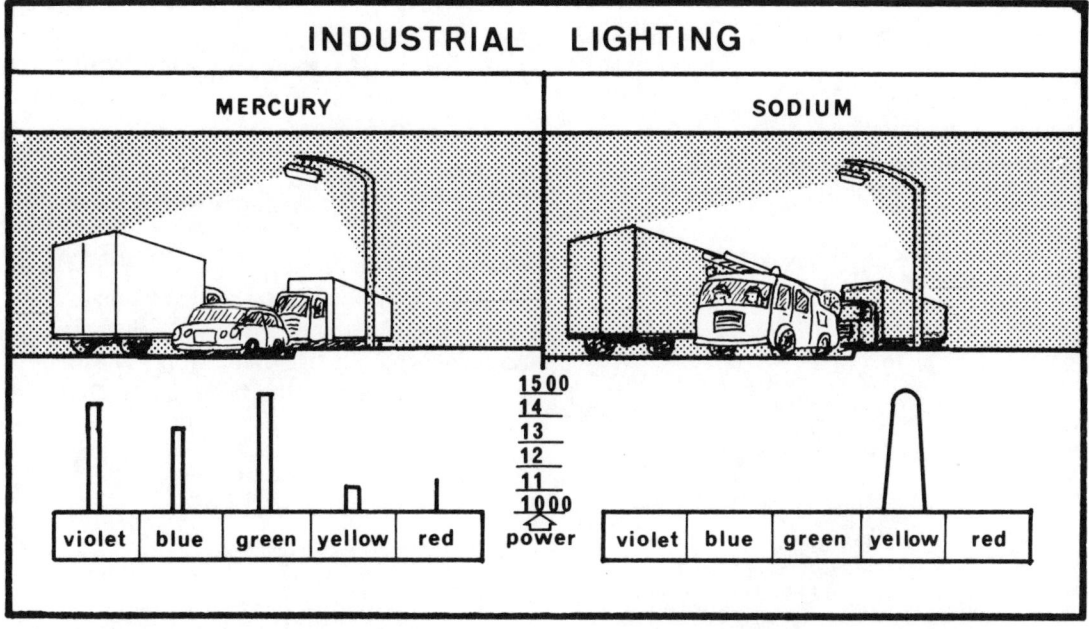

INDUSTRIAL LIGHTING

MERCURY

Visible spectrum consists of bands of colour - yellows and reds appear dull.

SODIUM

Only yellow can be seen clearly, all other colours appear brown or black

Fig. 14.4

Table 8 **Effects of artificial light on colour**

Colour of surface in daylight	Street lights		Domestic lighting	
	Sodium	Mercury	Tungsten	Fluorescent (natural)
Red	Brown	Brown or black	Bright red	Dull cool red
Blue	Brown or blue	Deep violet	Dull green-blue	Bright blue
Green	Brown-yellow	Dark green	Yellow-green	Cool blue green
Yellow	Yellow	Green yellow	Intense yellow	Green yellow
White	Light yellow	Blue white	Cream off-white	White

14.5 What is used to measure light intensity?

A light meter.

A light meter contains light sensitive material such as selenium which produces an electrical current when light falls on it (Fig. 14.5). The more light that falls on it the more electrical current is produced. The strength of the electrical current is indicated by a needle which moves across a scale on the meter. The intensity of the light is read from the scale.

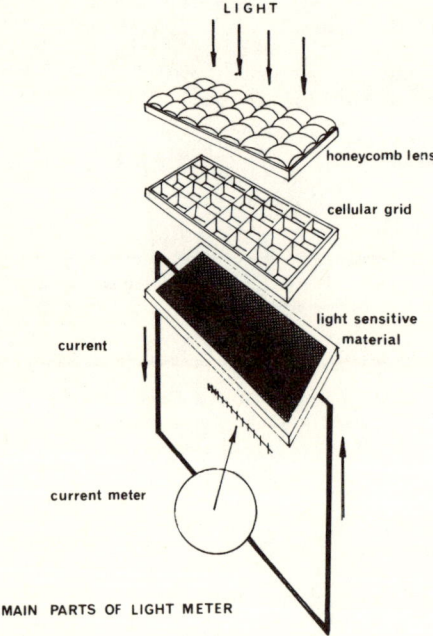

Fig. 14.5

Light meters are used to measure the intensity of daylight entering a room through a window, the level of artificial light emitted from a lamp, or the light reflected from painted walls. They assist architects and designers in deciding what lighting a building needs.

They are used also by photographers so that they know how wide to open the aperture of a camera to ensure that the film absorbs enough light to develop it. Cricket umpires also use them to decide whether the light is good enough to continue play.

14.6 In what way does the colour of walls affect the brightness of a room?

Coloured materials either reflect light and appear brighter, or absorb light and appear darker.

Colours can be measured in tones. The illustration shows a range of grey tones from white to black (Fig. 14.6). Those at the top are light tones and reflect most of the light, and those at the bottom are dark tones and absorb the light. These are REFLECTANCE VALUES of colours. If a room is

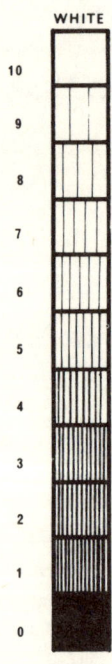

Fig. 14.6

painted in a colour similar in tone to number nine on the scale, most of the light will be reflected and the room will be bright. If a room is painted in a colour of number two tone, most of the light will be absorbed and the room will be dark.

Strong colours are often less reflective than dull looking colours. A post office red is a bright colour but has a reflectance value of four. An airforce grey is a dull colour but has a reflectance value of six. If bright red is used on large areas of a room it will absorb a lot of the light and the room will seem darker. Although airforce grey may seem to be a dark colour it will reflect more light than red and make a room seem less dark.

Rooms decorated with dark colours will need to have more, or brighter, lights than those with light colours.

14.7 How does light pass through some materials but not others?

Because some materials absorb less light than others.

Light is a form of ENERGY. Some materials absorb all of this energy, others absorb very little. Glass and some plastics absorb very little light. These materials are used to make windows, lamp lenses and anything else where it is necessary that light can pass through for the purpose of seeing. Because some light is always absorbed, the thicker the glass the less light gets through (Fig. 14.7A). Obviously it is better to keep the glass as thin as is practicable. White light is made up of many colours, which are called the SPECTRUM. By putting certain chemicals in when glass is produced a particular glass can be made to let only one colour through. This is a FILTER. All the other colours, which are different kinds of light energy, are absorbed by this glass. Light which has passed through the filter now is only of one colour.

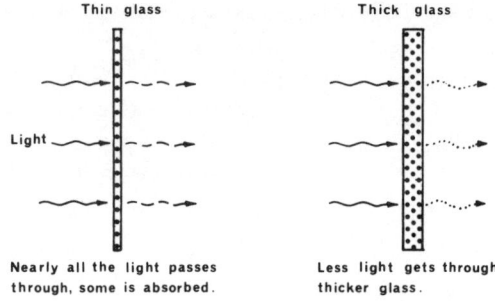

Thin glass — Nearly all the light passes through, some is absorbed.

Thick glass — Less light gets through thicker glass.

Fig. 14.7A

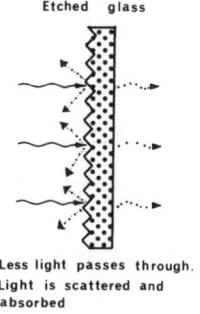

Etched glass

Less light passes through. Light is scattered and absorbed

Fig. 14.7B

Photographers may use filters over their lenses to reduce glare when photographing snow or sea scenes. Coloured spot lights are tungsten bulbs shone through a filtered lens.

Materials which appear black are absorbing all the colours. In other words, they absorb all the light energy.

If nearly all the light passes through a material it is called TRANSPARENT. Glass is transparent.

Some materials absorb more light than transparent materials but still let some through. These are called TRANSLUCENT. Varnishes and scumbles absorb some light depending upon the type of resin used, but some light does pass through so that the surface beneath can be seen. Varnishes are translucent.

Materials which absorb all the light are OPAQUE. Paint is opaque. The pigment absorbs or reflects the light. A mirror is glass which has been made opaque by coating one side with a silvery material that not only stops light passing through but reflects it. Opaque glass, which lets no light through, is available and some buildings are clad with it.

Some glass can be made translucent by texturing the surface. It can be moulded into a rough pattern so that, although light passes through, the image seen is diffused. The surface can be etched with acid or severely scratched by shotblasting which makes the glass absorb more light and become translucent (Fig. 14.7B).

LIFTING AND FIXING

15.1 How can heavy objects be moved easily?

15.2 How does the lever work?

15.3 Does the lever make building work easier and safer?

15.4 Are all pulley systems the same?

15.5 How are pulleys used in building work?

15.6 What is the screw?

15.7 How is the screw used in building work?

15.8 How does an inertia anchor work?

15.1 How can heavy objects be moved easily?

By the use of the lever and the pulley.

By using a strong iron bar as a lever, a person can push down on one end and move a load many times his or her own weight. By using ropes and pulleys a person can raise a load many times his or her own weight by a downward effort.

Fig. 15.1A

The lever and the pulley have MECHANICAL ADVANTAGE (MA). Mechanical advantage is calculated by dividing the load moved by the effort needed to move it.

If a lever is used to move a 150 kg weight by a 50 kg weight person or force, its MA is $\frac{150}{50} = 3$ (Fig. 15.1A). If the same person or the same force is applied to a pulley to raise a 100 kg weight, the pulley will have an MA of 2.

In all devices with MA the effort has to move further than the load.

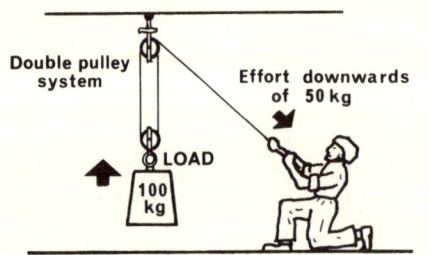

Fig. 15.1B

To move the weight on the lever, the effort will travel a long way downwards just to move the load a little distance (Fig. 15.1B).

With the pulley system, the person will have to haul down a long way to raise the load a short distance. If the MA of a lever is four then the effort will have to move four times as far as the load will move.

15.2 How does the lever work?

A lever is a rigid bar or rod which can be used to move or balance heavy loads.

At one end of the lever is the LOAD. At the other end is the EFFORT. A lever needs a point around which to pivot. This point is called the FULCRUM (Fig. 15.2A).

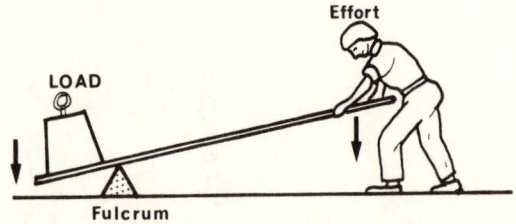

Fig. 15.2A

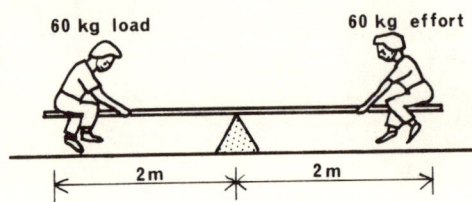

Fig. 15.2B

With the fulcrum in the middle of the lever and both the load and effort identical, the lever is balanced, or in EQUILIBRIUM (Fig. 15.2B).

If the position of the fulcrum is changed, the lever will move like a see-saw. One end will go up and the other will go down (Fig. 15.2C).

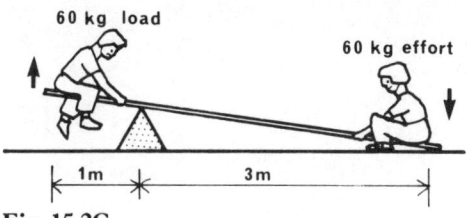

Fig. 15.2C

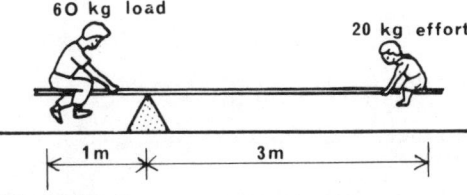

Fig. 15.2D

To balance a see-saw lever again without changing the position of the fulcrum it is necessary to change the amount of the effort. In this way a load of 60 kg can be balanced by an effort of only 20 kg (Fig. 15.2D).

This is how levers can be used to advantage. By positioning the fulcrum nearer the load than the effort, less effort is needed to balance the lever. However, the effort will travel a long way just to move the load a short distance.

15.3 Does the lever make building work easier and safer?

Yes, because of the MECHANICAL ADVANTAGE obtained from levers. This advantage enables people to move large loads and exert high pressures without injuring themselves, and with less risk to others.

The principle of the lever is used in the design of many tools and pieces of equipment. A claw-hammer or a pair of pincers act as levers when used to pull nails out of timber (Fig. 15.3A). A spanner is a lever when used to tighten nuts (Fig. 15.3B).

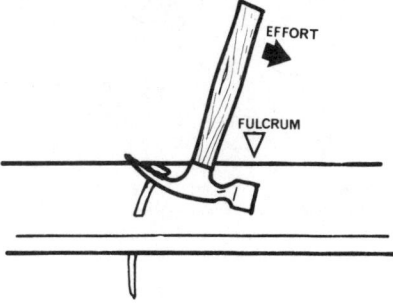

Fig. 15.3A

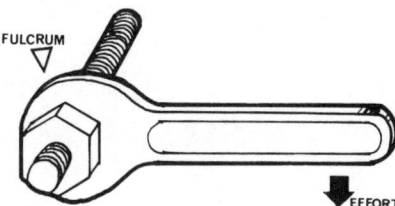

Fig. 15.3B

Although the most common lever has the FULCRUM placed between the LOAD and the EFFORT, these positions are varied on other forms of lever.

On a cradle the fulcrum is the point at which the outrigger overhangs the building (Fig. 15.3C). It is always very close to the load (the cradle).

The fulcrum on a wheelbarrow is the wheel. Provided the load is placed near the fulcrum the wheelbarrow is a very efficient lever (Fig. 15.3D).

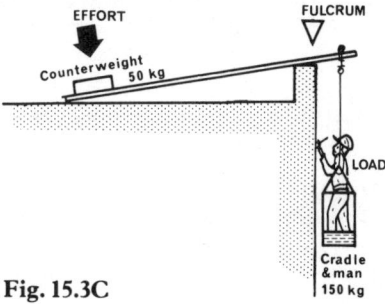

Fig. 15.3C

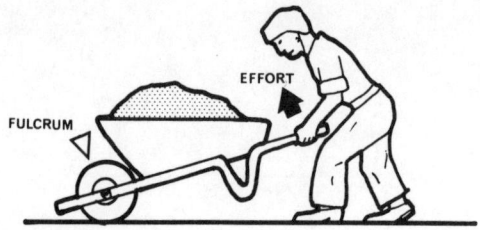

Fig. 15.3D

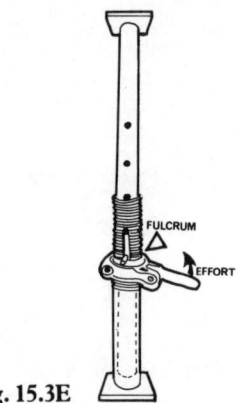

Fig. 15.3E

The handle of an *'acrow' prop* is a lever. It operates the rotating sleeve which moves along the threaded part. The fulcrum is the point at which the threaded parts meet (Fig. 15.3E).

15.4 Are all pulley systems the same?

No. A fixed pulley is the simplest and least efficient. Movable pulleys are more elaborate and more efficient.

A fixed pulley has one fixed pulley block with a single wheel. It makes the job of moving small loads very easy and convenient, by changing the direction of force. Instead of a man straining *up* to lift a load he can heave *down* and use his own body weight and muscles to a greater advantage. The

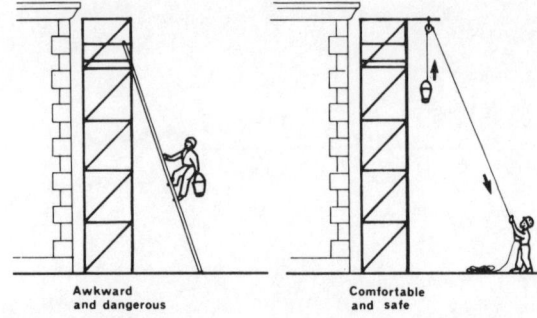

Fig. 15.4A

effort goes *down* and the load goes *up* (Fig. 15.4A).

Movable pulleys have one fixed and one movable pulley block, some with more than one wheel in them (Fig. 15.4B).

Movable pulleys have a MECHANICAL ADVANTAGE (MA) and enable loads to be moved with less effort. A rough guide to the MA gained is by counting the number of wheels in the pulley system, or the number of ropes supporting the load. If, for example, there are three, the system has an MA of three. This means that a man weighing, say, 80 kg could heave down with all

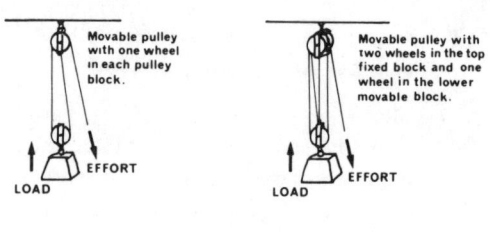

Movable pulley with one wheel in each pulley block.

Movable pulley with two wheels in the top fixed block and one wheel in the lower movable block.

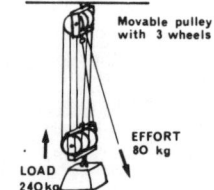

Movable pulley with 3 wheels

Fig. 15. 4B

his weight and lift a load of $3 \times 80\,kg = 240\,kg$. He would, however, move the load only a short distance. For every metre the load moves the effort will have to move three metres.

15.5 How are pulleys used in building work?

Pulley systems are used to raise and lower loads with much less effort.

Pulley systems are also known as pulley blocks or blocks and tackle. They are threaded together with rope, wire rope or chains.

A gin wheel is an example of a fixed pulley. It is limited to small loads that weigh less than the average man, say 50 kg. The man pulls *down* on the rope and the load goes *up*. It is a safe way of moving small loads (Fig. 15.5A).

Movable pulleys are used to raise loads which weigh more than the average man.

A safety or bosun's chair is suspended on a movable pulley system. By hauling down on the

Fig. 15.5B

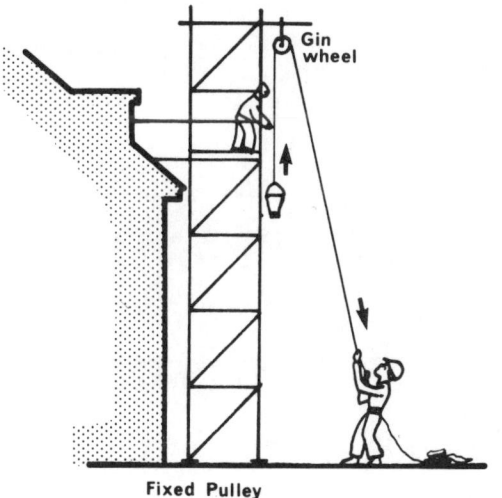

Fixed Pulley

Fig. 15.5A

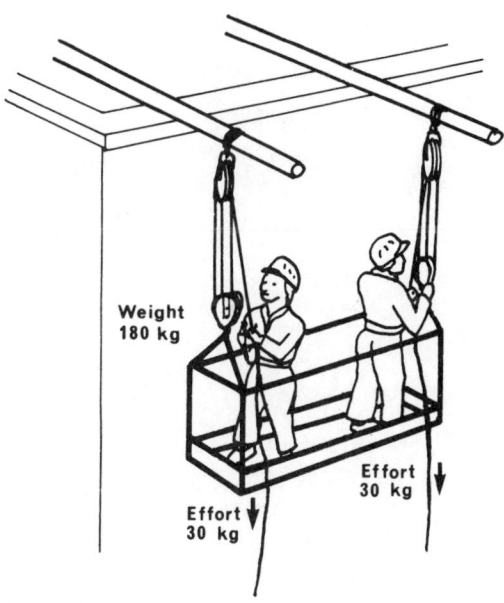

Fig. 15.5C

free rope, a man can raise himself safely and with ease. Although the man and chair may weigh, say, 90 kg he has to exert a force of only half that to move the chair (Fig. 15.5B).

Cradles have movable pulleys which reduce the effort needed to raise the cradle to only a third of the load. The combined weight of the cradle and men is spread evenly between the two pulley systems. If the combined weight is 180 kg, each pulley takes 90 kg. To raise one side of the cradle a man will have to exert a force of $\frac{90}{3} = 30$ kg (Fig. 15.5C).

15.6 What is the screw?

The screw is a continuous groove cut around a cylinder. The continuous groove is called a thread and by rotating the screw against a similar thread, movement to-and-fro is produced (Fig. 15.6A).

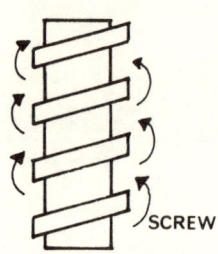

Continuous 'ramp' along which another 'thread' can move

Fig. 15.6A

Movement is produced because the thread on the screw acts as a continuous ramp or sloping platform, just like the slide part of a helter-skelter (Fig. 15.6B). When interlocked with a similar thread, and rotated, the effort moves one thread along the ramp produced by the other.

When threads are interlocked, they cannot be pulled apart, so they resist pressures which may

pull at them. They will remain together provided the threads do not break or SHEAR (Fig. 15.6C, D).

There is more resistance to shearing when many small threads are interlocked than with just a few large or coarse threads. Pressures which may pull at the threads are shared (Fig. 15.6E).

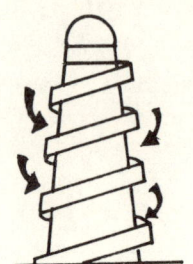

Fig. 15.6B Continuous 'ramp' on a helter skelter

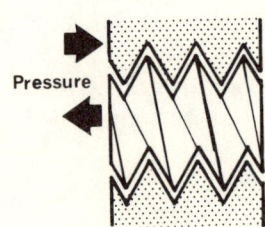

Pressure

Fig. 15.6C Threads interlocked

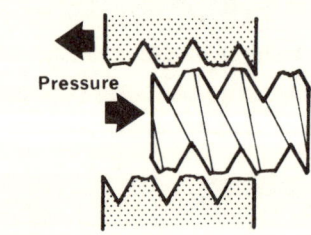

Pressure

Fig. 15.6D Threads sheared

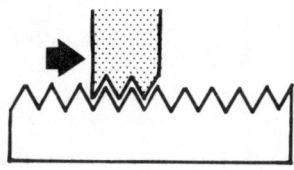

Pressure shared by only
3 threads - weak bond

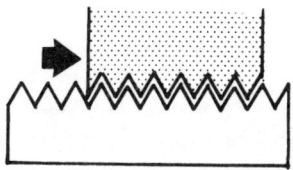

Pressure shared by
7 threads - strong bond

Fig. 15.6E

15.7 How is the screw used in building work?

It gives the movement and fixing strength to wood screws and machine screws, or bolts, which provide the main jointing methods in construction work. It also gives the movement in the screw operated jacks, such as props, and the clamping movement in equipment like vices, drill chucks, and reveal ties.

When threads are interlocked, and rotated, movement to-and-fro is produced. Movement is obtained in a number of ways (Fig. 15.7A).

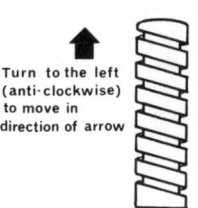

Turn to the left
(anti-clockwise)
to move in
direction of arrow

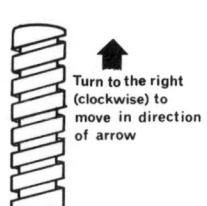

Turn to the right
(clockwise) to
move in direction
of arrow

Fig. 15.7A

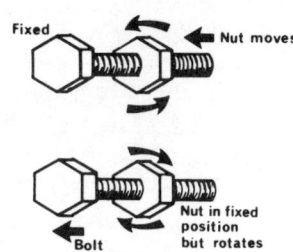

Fixed   Nut moves

Nut in fixed
position
but rotates

Bolt
moves

Fig. 15.7B

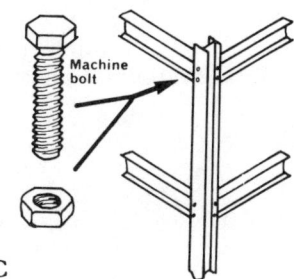

Machine
bolt

Fig. 15.7C

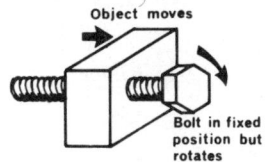

Object moves

Bolt in fixed
position but
rotates

Fig. 15.7D

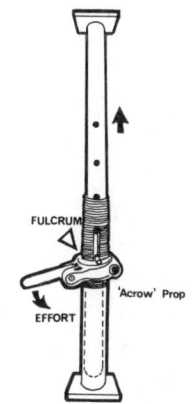

FULCRUM

'Acrow' Prop

EFFORT

Fig. 15.7E

A threaded nut will move along a threaded bolt when rotated. By applying extra pressure with a lever, in the form of a spanner, a very tight fit can be obtained. Structural steel girders are joined in this way (Figs. 15.7B, C) and screw operated jacks are extended (Fig. 15.7E).

If the nut rotates, but is held in fixed position, the bolt moves.

Reveal ties, adjustable base plates and adjustable spanners work in this way (Fig. 15.7F).

A screw moves when rotated in a threaded hole, which is in a fixed position (Fig. 15.7G). A drill chuck clamps a drill with wedged jaws which

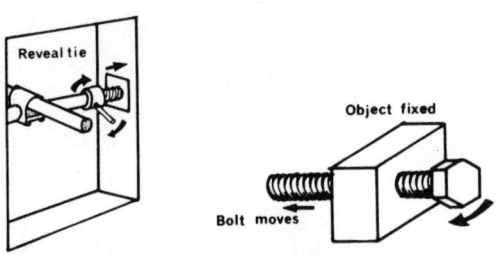

Fig. 15.7F   **Fig. 15.7G**

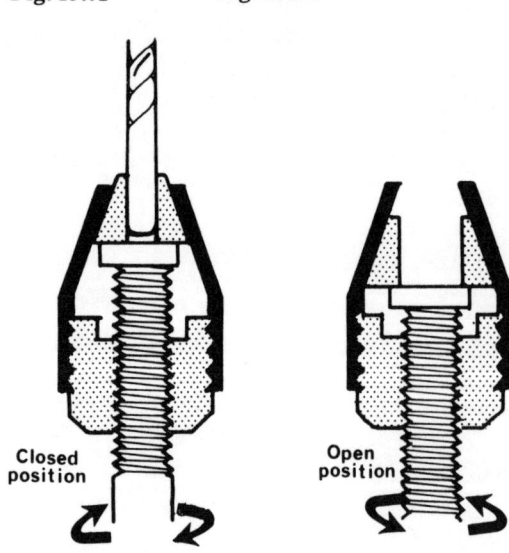

Fig. 15.7H

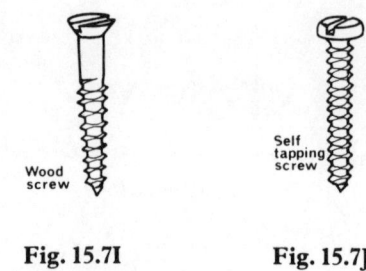

Fig. 15.7I   **Fig. 15.7J**

close together when forced up against the sloping sides of the chuck by a screw (Fig. 15.7H).

Wood screws are used to join wood and similar soft materials (Fig. 15.7I). They are cut with a sharp tapered thread, finishing in a point. When turned into a suitable sized hole, the screw cuts a thread into the sides of the hole and moves down into it.

Self tapping screws have parallel sides and are used to join metal (Fig. 15.7J). They are made from hardened steel and have a sharp thread which can cut a thread into the sides of a hole drilled into metal.

15.8 How does an inertia anchor work?

A sudden downward movement of the anchor makes it jam tight on the fixed cable. The inertia of movable steel balls inside the anchor produces the jamming effect (Fig. 15.8A).

INERTIA means reluctance to move. The bigger something is, the harder it is to get it moving. When a bus suddenly moves away from a stop, for a fraction of a second the passengers remain stationary. Then they feel a *push in the back* from the seat.

Inside an inertia anchor are three steel balls which are free to move inside a cone shaped container. When the anchor moves slowly along the fixed cable the balls roll freely. Should the anchor suddenly move downwards at a fast rate the three balls are left stationary, just like the passengers on the bus. The anchor moves but the balls remain stationary by their inertia. This means that the balls move into the narrow part of

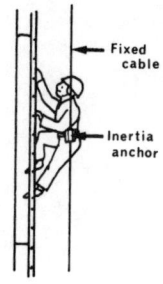

Fig. 15.8B

the conical container and jam against the fixed cable. Inertia reel safety belts in cars operate in a similar way. During a crash, the sudden movement forward of the passenger makes the belt jam. As with inertia anchors on safety lines, while the anchors are moved slowly the seat belts can be pulled forward easily (Fig. 15.8B).

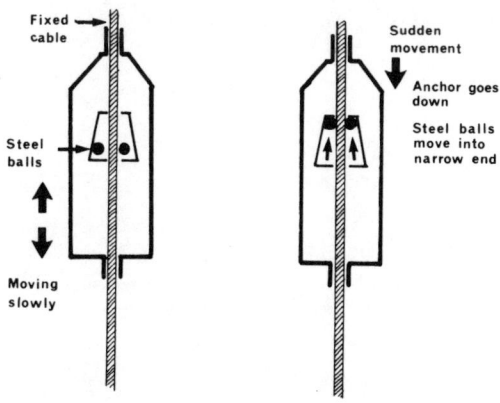

Fig. 15.8A

131

PUMPS AND PRESSURES

16.1 What is a siphon?

16.2 What is suction?

16.3 What is atmospheric pressure?

16.4 What is a pump?

16.5 How do pumps work?

16.6 Which pumps are used in building work?

16.7 What is hydraulic pressure?

16.8 How is hydraulic pressure used?

16.9 What is a venturi?

16.1 What is a siphon?

A device which enables a fluid to be transferred from a high level to a low level over an obstacle.

Just as water flows naturally down hill by the force of gravity, so the siphon works by the same principle. To start a siphon the liquid must be pulled up and over the obstacle, and down to a point lower than the surface of the liquid. This is usually done with a tube. When there is more liquid in the longer length of tube outside the container than in the shorter length inside gravity makes the liquid flow out. It continues to flow because MOLECULES of liquids hold on to one another (Fig. 16.1A).

Fig. 16.1B

moves out of the glass to a lower level (Fig. 16.1B).

The most common application of the siphon is the flushing system of a lavatory cistern or water closet (Fig. 16.1C). An inverted U-shaped tube is connected to the flush pipe. When the tube is filled with water by pulling the handle, the full tank of water is flushed down to the pan by siphonic action.

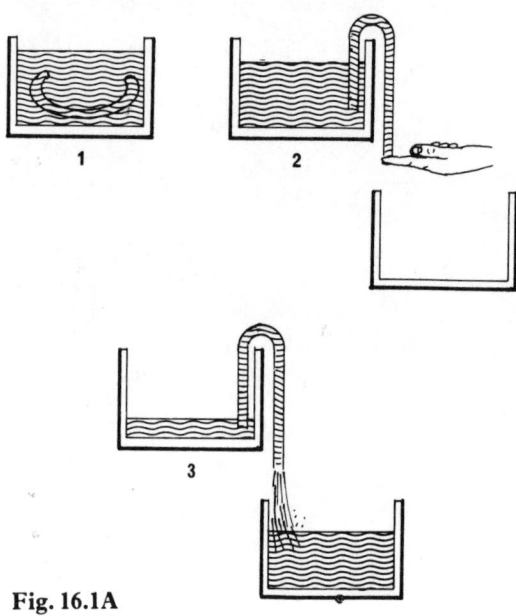

Fig. 16.1A

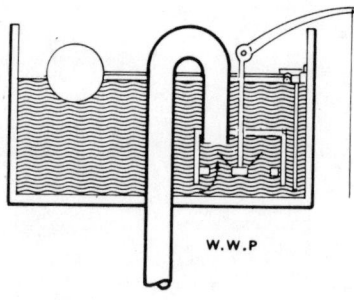

Fig. 16.1C

16.2 What is suction?

Suction means the removal of air to make a partial VACUUM. Any force that is exerted comes from ATMOSPHERIC PRESSURE.

When a rubber sucker is pushed down air is squeezed out. It is the difference in pressure between the atmosphere pushing down and the pressure inside the rubber sucker which holds it to the surface. When drinking through a straw the cheeks are moved so that air pressure in the mouth is less than atmospheric pressure. The

The principle of the siphon can be shown by resting a fine chain, which is held together by its links, in a glass, with the end of the chain hanging over the edge of the glass to a point just below its base. When it is released the rest of the chain

difference in pressure forces the liquid up the straw and into the mouth.

Suction pumps cannot lift a liquid any higher than atmospheric pressure will push it.

16.3 What is atmospheric pressure?

The force created by billions of tiny particles of the atmosphere bouncing off our bodies.

Objects on the surface of the earth are surrounded by air. This is the atmosphere and most of it is found within 80 km of the earth's surface. The atmosphere is a mixture of gases.

Gases consist of tiny particles known as MOLECULES which are moving about very fast and bouncing off each other and other objects. Billions and billions of these molecules are bouncing off the surface of our bodies and other objects all of the time. This makes a constant pressure on us. We do not feel this pressure as we have air in the cells of our bodies that pushes back with an equal force.

An example of what happens when there is not an equal force pushing out against atmospheric pressure is when all the water is drawn out of a hot water cylinder and no air can get in to replace it. Atmospheric pressure crushes the cylinder.

The pressure of the atmosphere at sea level is roughly 10 000 kg per square metre (15 pounds per square inch). Air is most dense at sea level and gets steadily less dense as height increases. This is because air is squashed, or compressed, by the weight of the air above it. As a mountaineer climbs higher, so the air gets *thinner*, so there are less molecules to rebound from him.

ATMOSPHERIC PRESSURE on earth varies slightly according to the weather. High pressure usually means fine weather and low pressure means bad weather.

16.4 What is a pump?

A machine which moves liquids or gases.

A simple example of a pump is a bicycle pump. Other machines which pump air are called compressors because they squash the air or other gas into a smaller space.

Pumps are hand operated or worked by an engine or an electric motor. The operating mechanism of the pump moves either to-and-fro, and is called a *reciprocating pump* (Fig. 16.4A), or round and round and is called a *rotary pump* (Fig. 16.4B). The pump draws liquid in at an *inlet* and pushes it out at an *outlet*.

A *reciprocating pump* can be used for an airless spray unit (Fig. 16.4C).

With an electrically operated air compressor an electric motor drives the compressor (Fig. 16.4D). Inside the compressor a piston moves up

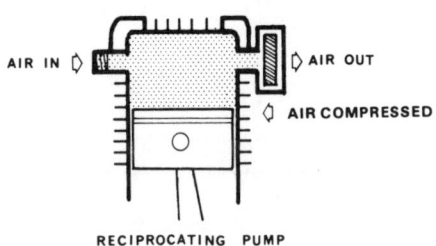

RECIPROCATING PUMP

Fig. 16.4A

ROTARY PUMP

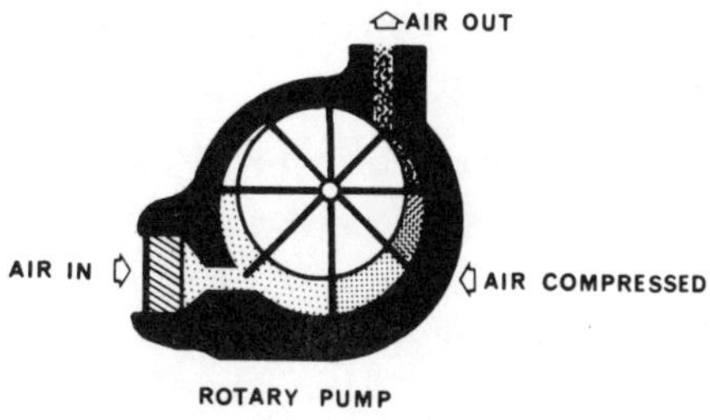

Fig. 16.4B

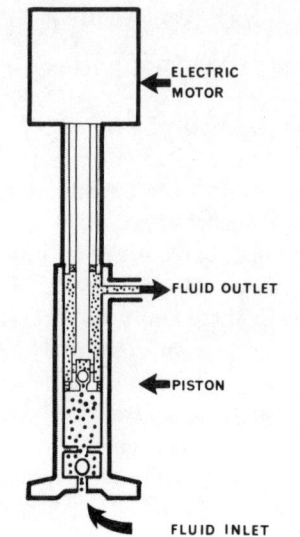

Fig. 16.4C

ELECTRIC MOTOR

FLUID OUTLET

PISTON

FLUID INLET

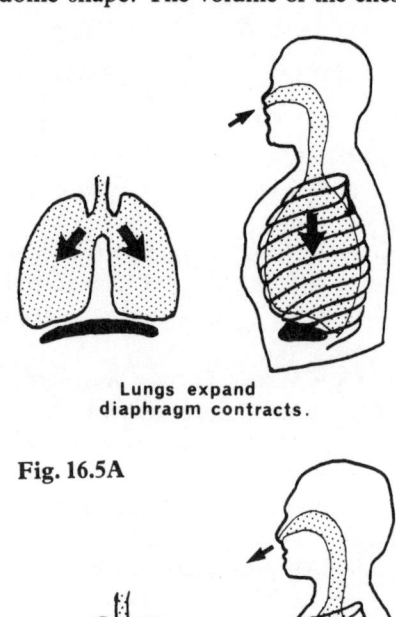

liquid into the pump's mechanism. The mechanism then pushes the gas or liquid to wherever it is required.

The principle of a pump is similar to that of the human diaphragm. The diaphragm is a sheet of muscle attached to the ribs. It is dome shaped. When the diaphragm contracts it flattens and the chest gets bigger. Because the lungs are inside the rib cage they also get bigger and the pressure of air is lower inside them than outside the body. So air moves into the lungs to even the pressure. When we breathe out the diaphragm returns to the dome shape. The volume of the chest cavity

Lungs expand
diaphragm contracts.

Fig. 16.5A

1 COMPRESSOR	**5** OUTLET
2 DRIVE BELT	**6** AIR RECEIVER
3 ELECTRIC MOTOR	**7** INLET
4 TRANSFORMER	

Fig. 16.4D

and down. As it goes down it sucks in air. As it goes up it pushes the air into a storage tank.

16.5 How do pumps work?

By creating a partial VACUUM inside a casing. This allows ATMOSPHERIC PRESSURE to push the gas or

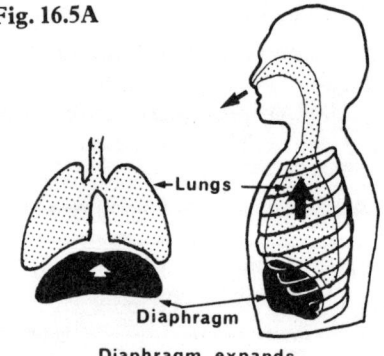

Lungs

Diaphragm

Diaphragm expands
lungs contract.

Fig. 16.5B

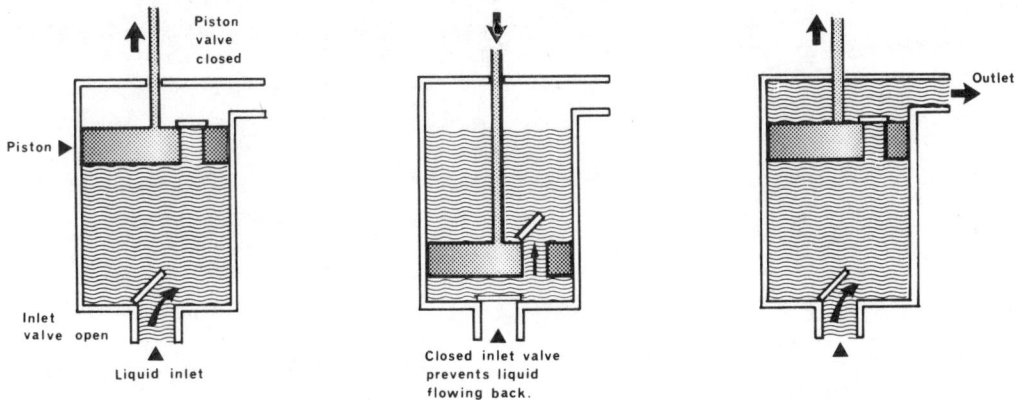

SUCTION PUMP or LIFT PUMP

Fig. 16.5C

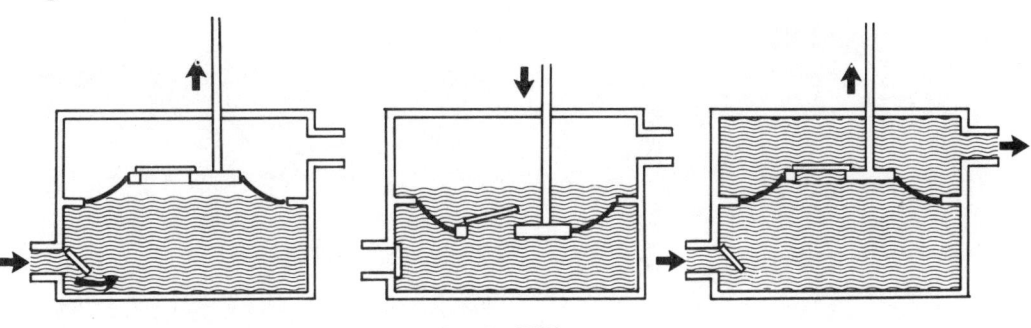

DIAPHRAGM PUMP

Fig. 16.5D

gets smaller and so the pressure of air in the lungs builds up and becomes greater than the pressure outside. This forces air out of the lungs (Fig. 16.5A, B).

A simple type of pump is the SUCTION or LIFT PUMP. It contains a piston and two valves. Valves allow the liquid to pass one way only. As the piston rises, its valve closes and a partial vacuum is created beneath. Liquid is forced into the chamber by atmospheric pressure. As the piston falls it passes through the liquid. As the piston rises a second time it lifts the liquid to the outlet. At the same time, more liquid is *sucked* in below and the cycle of *sucking in* and *pushing out* continues (Fig. 16.5C).

Other types of pumps are diaphragm, centrifugal and gear pumps.

Diaphragm pumps (Fig. 16.5D) use a flexible diaphragm instead of a piston. It is made of rubber or plastic and moves up and down very quickly. The action is similar to that of the human diaphragm.

Centrifugal pumps (Fig. 16.5E) use rotating impellers, which are like fans. Liquid is directed at the impellers which deflect it and send it off in the direction that the vanes are angled. As it changes direction the liquid's movement creates a pressure drop. Liquid then moves in at the inlet to even up the pressure.

Gear pumps use rotating interlocking gear

137

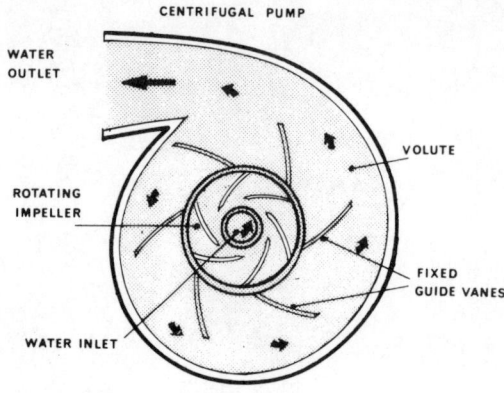

CENTRIFUGAL PUMP

WATER OUTLET

VOLUTE

ROTATING IMPELLER

FIXED GUIDE VANES

WATER INLET

Fig. 16.5E

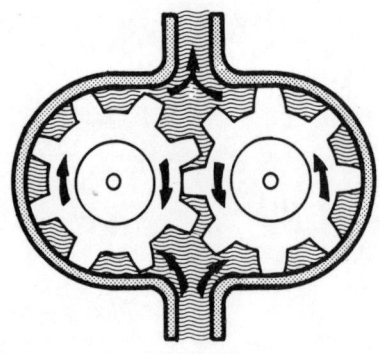

ROTARY GEAR PUMP

Fig. 16.5F

wheels (Fig. 16.5F). The rotating of the wheels moves the liquid faster and more liquid moves in to keep the pressure even. The meshed gears prevent the liquid running back.

16.6 Which pumps are used in building work?

Pumps are used to supply paint to the spray gun in airless spray equipment (Fig. 16.6A). This is a *lift pump* and is often operated by an air motor. Other types operate with a *diaphragm pump* driven by an electric motor.

Gear pumps are used to circulate lubricating oil around engines.

Surface water from trenches and excavations is moved by a *diaphragm pump* (Fig. 16.6B). If sand or stones are sucked into this type of pump the diaphragm would not be damaged. A piston pump would suffer a lot of damage.

Centrifugal pumps are used to circulate hot water in a heating system.

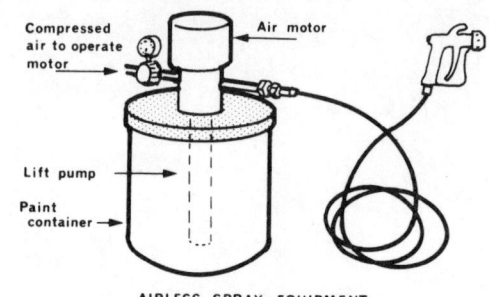

Compressed air to operate motor

Air motor

Lift pump

Paint container

AIRLESS SPRAY EQUIPMENT

Fig. 16.6A

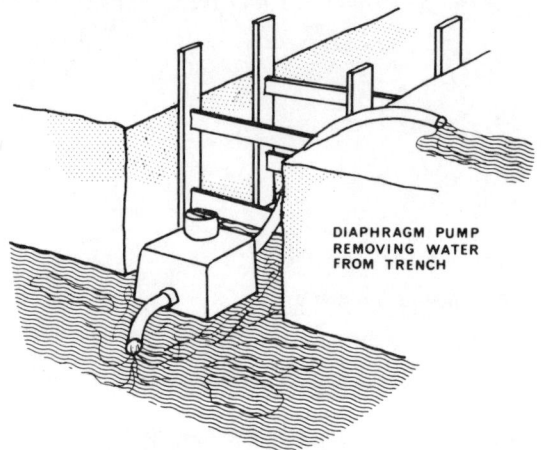

DIAPHRAGM PUMP REMOVING WATER FROM TRENCH

Fig. 16.6B

16.7 What is hydraulic pressure?

It is a term used when referring to pressures in liquids.

Liquids like water do not compress very easily, so they can be used to send a force from one place to another (Fig. 16.7A). Air squashes when any pressure is put on it so air cannot be used in the same way.

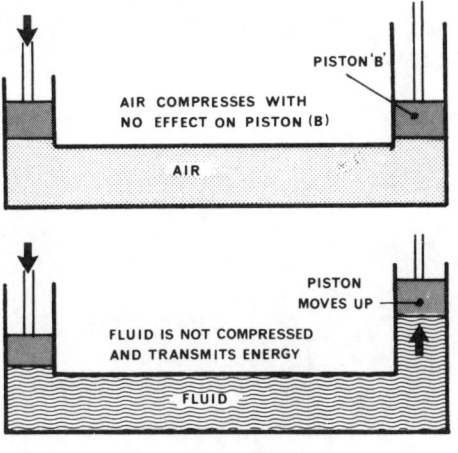

HYDRAULIC PRESSURE

Fig. 16.7A

If a piston is pushed *down* on one end of a container of liquid the force is transmitted through the liquid to the other end of the container where another piston will be moved *up*. Car braking systems use this method. It is easier to run a pipe containing a liquid around to each of four wheels than to arrange a number of levers, rods and linkage which may fail. Pressure pushed on one end of the pipe is transmitted equally to each of the four outlets attached to the brakes (Fig. 16.7B).

Care must be taken with HYDRAULIC systems to ensure that there are no air bubbles in the liquid. These bubbles will compress and the force would not be sent to where it is needed. In a car these air bubbles make the brakes feel *spongy* and they do not work efficiently. In such cases the pipes must be bled to get rid of the air.

HYDRAULIC PRESSURE systems are used also in car jacks and lifting devices. A small force at one end, moved over a long distance, pushes fluid into a larger container. The level in the larger container rises slightly and lifts a much heavier weight a smaller distance.

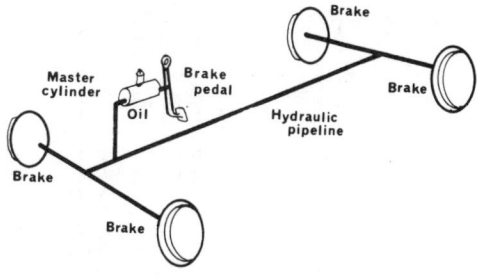

Fig. 16.7B

16.8   How is hydraulic pressure used?

To lift or push loads and to act like a cushion in shock absorbers. It is also used to pump liquids.

A piston moving against a liquid meets a lot of resistance if the liquid has only a small hole through which to escape (Fig. 16.8A).

Shock absorbers use hydraulic pressure in this way. A piston moves up and down inside a cylinder and forces oil through narrow holes or valves. This slows down the movement and

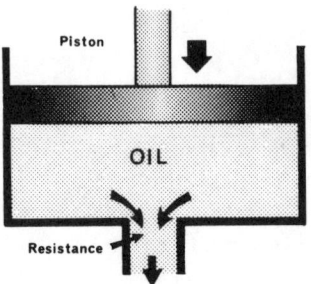

Fig. 16.8A

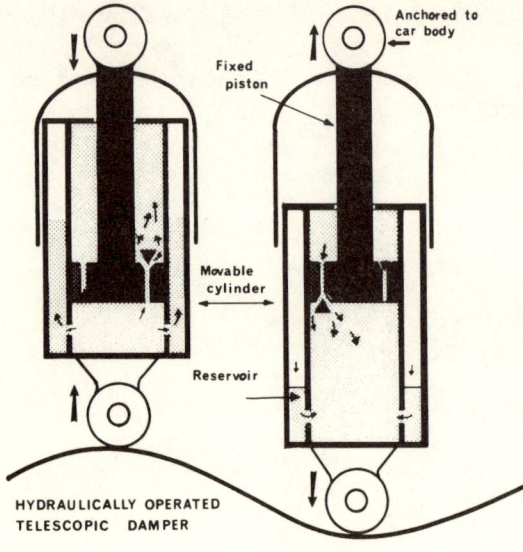

Fig. 16.8B

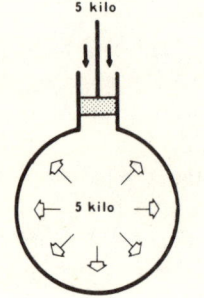

Fig. 16.8C

quickly. A water pistol and hypodermic syringe work in this way.

If the piston is driven by a motor then the speed and power of the piston can be increased very much and the pressure on the liquid is increased. Most fluid pumps used to move liquids are based on this principle, such as airless spray pumps and water pumps.

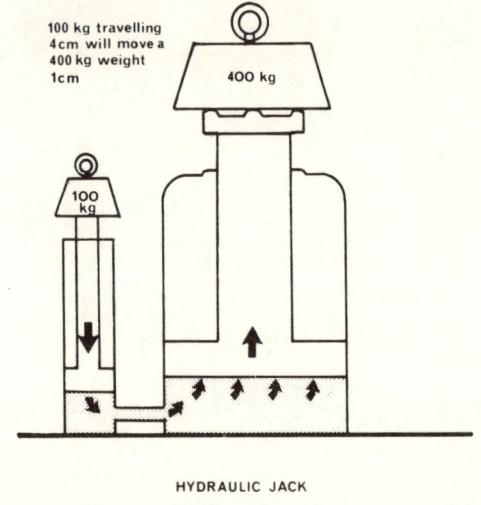

Fig. 16.8D

16.9   What is a venturi?

A tube which gets narrower in one part.

makes the impact softer. When the pressure is off, the oil returns into the first container and waits for another push or pump (Fig. 16.8B).

When pressure is applied to a liquid in a container the same pressure is sent through the liquid which pushes against every part of the surface with equal force (Fig. 16.8C).

Hydraulic jacks lift weights like this (Fig. 16.8D). A similar principle is used to operate hydraulic brakes in cars and lorries.

If a liquid which is under pressure is allowed to escape through a small hole it is forced out very

If air passes through a tube which suddenly gets narrower and then widens, its speed increases and its pressure drops. This fact can be used in many ways.

Carburettors are used in petrol engines to mix petrol and air. The VENTURI is situated above the petrol supply (Fig. 16.9A). Inside the venturi the pressure of air is lower than ATMOSPHERIC PRESSURE. This difference causes the petrol to be pushed up through the jet and into the venturi where, in the form of a fine spray, it is mixed with the air. This mixture is then sent to each cylinder in turn to be burnt. The venturi effect is used on a suction feed spray gun to push the paint up into the air stream (Fig. 16.9B). It is also used in the nozzle of open blast cleaning equipment.

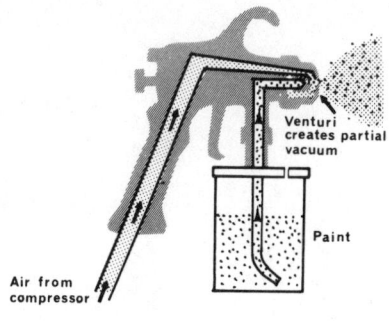

VENTURI

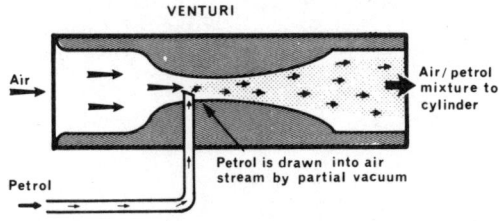

Air

Air/petrol mixture to cylinder

Petrol is drawn into air stream by partial vacuum

Petrol

Fig. 16.9A

Venturi creates partial vacuum

Paint

Air from compressor

Fig. 16.9B

GLOSSARY

Words in CAPITALS denote cross-references with other key words in the text.

ABSOLUTE ZERO – The lowest temperature theoretically possible. Zero degrees KELVIN; minus 273 degrees CELSIUS.

ABSORBENCY – Generally, the ability of a substance to (a) take in something else, e.g. moisture, or (b) to reduce the intensity of something else, e.g. sound, light, heat. See CAPILLARITY; INSULATION.

ABSORBENT – A substance is absorbent if it possesses ABSORBENCY.

ACCELERATOR – Substance which increases the rate of a CHEMICAL REACTION. Some are OXIDIZERS. See CURING; POLYMERIZATION; RETARDER.

ACID – Compound containing hydrogen which is replaced by a metal during NEUTRALIZATION to form a SOLUTION of a SALT. ACIDIC in its properties.

ACIDIC – Having the properties of an ACID, i.e. turns blue LITMUS red, pH VALUE between 1–6, NEUTRALIZED by an ALKALI when in SOLUTION to form a SALT. Weak acids have a sour taste. Strong acids are highly corrosive.

ACOUSTICS – The study of sound. The quality, or audibility of sound, in a room etc. See ECHO; REVERBERATION; VOLUME.

ADDITION POLYMER – POLYMER produced by a CHEMICAL REACTION involving the joining together of many MOLECULES to form much larger MOLECULES. See CONDENSATION POLYMER.

ADHESION – Attraction between MOLECULES of different substances. Force which 'sticks' a substance to a surface. See COHESION.

ADHESIVE – (1) Possessing powers of ADHESION. (2) Substance used for sticking surfaces together, i.e. glue.

AIR DRYING – Air drying coatings are those which HARDEN at normal room temperature, as opposed to 'forced drying' coatings which require HEAT to CURE.

AIR SET – Plasters and cements air set when they partially HYDRATE by absorbing WATER VAPOUR from the air. See HYDRATION.

ALKALI – Most commonly the soluble hydroxide of a metal. Neutralizes ACIDS to form a SOLUTION of a SALT. ALKALINE in its properties.

ALKALINE – Having the properties of an ALKALI, i.e. turns red LITMUS blue, pH VALUE between 8–14, neutralizes ACIDS when in solution to form SALTS. Strong alkalis are caustic, i.e. corrosive to ORGANIC matter.

ALLOY – Mixture of two or more metals, or sometimes a metal and a non-metal e.g. brass (copper and zinc), stainless steel (steel, chromium and nickel).

AMORPHOUS – Non-CRYSTALLINE, having no definite shape or form.

AMPERE. (amp) – Unit of quantity of ELECTRIC CURRENT. The quantity of electricity flowing through a CONDUCTOR. Amps = Watts divided by Volts. See WATT; VOLT.

ANODE – Positive ELECTRODE in ELECTROLYSIS. In GALVANIC CORROSION, the metal which is decomposed by the ELECTRIC CURRENT. See CATHODE.

ANODIZING – Forming an OXIDE film on aluminium or other light ALLOY by ELECTROLYSIS.

ATMOSPHERIC POLLUTION – Carbon, sulphur and other chemicals, and various dusts which may collect in the air in excessive amounts.

ATMOSPHERIC PRESSURE – The pressure exerted on the Earth by the weight of the surrounding air. Approximately 14.72 psi; 1 bar; 760 mm of mercury at sea level. $10\,000$ kg/m_2, 1 kg/cm_2.

ATOM – The smallest part of an ELEMENT which can take part in a CHEMICAL REACTION. See ELECTRON; MOLECULE.

ATOMIZATION – Breakup of a LIQUID into a spray of fine particles by pressure, usually by compressed air or HYDRAULIC PRESSURE.

BATTERY – A bank of two or more PRIMARY CELLS coupled together.

BI-METAL STRIP – Strips of two different metals fixed together face to face. When heated, DIFFERENTIAL EXPANSION causes the strip to bend. Used in some THERMOMETERS and THERMOSTATS.

BLEACH – (1) Liquid used to remove COLOUR from substances. Releases chlorine which acts as an OXIDIZER. (2) Generally, to remove or reduce colour.

B.T.U. (B.t.u. British Thermal Unit) – Old unit of quantity of HEAT. The amount of heat needed to raise the TEMPERATURE of 1 lb of water 1 degree Fahrenheit. Equal to 251 CALORIES, or 1055 JOULES.

CALORIE – Metric unit of quantity of heat. The amount of heat needed to raise the TEMPERATURE of 1 gm of water 1 degree CELSIUS. Equal to 4.2 JOULES.

CALORIFIC VALUE – See HEAT OF COMBUSTION.

CAPILLARY ACTION (Capillary attraction, Capillarity) – The action of LIQUIDS to rise in narrow tubes and gaps. Principle of wicks and blotting paper which are ABSORBENT and POROUS.

CARBON CHAIN – Term for long 'chain-like' MOLECULES of ORGANIC substances. Consists of a long chain of carbon ATOMS with other atoms attached.

CATALYSIS – CHEMICAL CHANGE of one substance by the addition of another, a CATALYST, which itself undergoes no change.

CATALYST – Substance which alters the rate at which a CHEMICAL REACTION occurs. Brings about a change by CATALYSIS.

CATHODE – Negative ELECTRODE in ELECTROLYSIS. In GALVANIC CORROSION, the metal which is given SACRIFICIAL PROTECTION. See ANODE.

CELSIUS (°C) – Metric unit of measurement of TEMPERATURE. Based on 100 divisions, zero being the melting point of ice, and 100 being the boiling point of water. Identical to the CENTIGRADE scale.

CENTIGRADE – See CELSIUS.

CHANGE OF STATE – Change from one STATE OF MATTER to another by either a PHYSICAL CHANGE or a CHEMICAL CHANGE. See EVAPORATION; CONDENSATION; LIQUEFIED; REVERSIBLE COATING; THERMO-PLASTIC; SOLUTION.

CHEMICAL CHANGE – A change in a substance which involves a CHEMICAL REACTION. As a result, a new substance is formed. See ACCELERATOR; POLYMERIZATION.

CHEMICAL COMPOUND – See COMPOUND.

CHEMICAL FORMULA – Shorthand which describes the atomic composition of one MOLECULE of a substance. See CHEMICAL NAME; CHEMICAL SYMBOL.

CHEMICAL NAME – Accurate descriptive name by which COMPOUNDS are known, as opposed to a variety of 'common' names. See CHEMICAL FORMULA.

CHEMICAL REACTION – A reaction which results in a re-arrangement of ATOMS to form a new product. See CHEMICAL CHANGE.

CHEMICAL SYMBOL – Symbolic representation of an ELEMENT consisting of its first letter, or its first and another letter of its name. Sometimes a foreign name is used.

COALESCENCE – (1) The joining-up of particles in an EMULSION caused by the attraction of one MOLECULE to another. (2) Union of CRYSTALS into a larger unit.

COHESION – Attraction between MOLECULES within a substance. Force holding together SOLIDS and LIQUIDS. See ADHESION.

COHESIVE – Possessing powers of COHESION.

COLLOID – See COLLOIDAL SOLUTION.

COLLOIDAL SOLUTION – State intermediate between a true SOLUTION and a SUSPENSION. A dispersion of very fine particles, so minute that they do not settle out, i.e. glue in water, varnish. See GEL; EMULSION.

COLOUR – The effect of LIGHT of a particular wavelength upon the eye. See ELECTROMAGNETIC ENERGY; SPECTRUM.

COMBUSTIBLE – Able to be set on fire. Usually related to SOLIDS which will burn. See FLASH POINT; FLAMMABLE; SPONTANEOUS COMBUSTION.

COMBUSTION – Burning. CHEMICAL REACTION between the substance burning and oxygen in the air. Heat, and often light and flames are produced. See COMBUSTIBLE.

COMPOUND – Substance made by joining together two or more ELEMENTS.

CONDENSATION – CHANGE OF STATE of a VAPOUR to a LIQUID by cooling, or by increasing the pressure. See CONDENSATION POLYMER; DEW POINT; SATURATED ATMOSPHERE.

CONDENSATION POLYMER – POLYMER produced a CHEMICAL REACTION during which small molecules link together to form a larger molecule and a simple molecule, e.g. water or alcohol is eliminated. See ADDITION POLYMER.

CONDUCTION – (1) The transmission of HEAT in a substance from a place of higher TEMPERATURE to a place of lower temperature without a movement of the mass of the substance. Heat energy is moved by collision of atoms or molecules with their neighbours. More pronounced in SOLID substances, especially metals. (2) The flow of ELECTRICITY along a CONDUCTOR. See INSULATION.

CONDUCTIVITY (Electrical) – The ability of a material to allow the flow of electricity. See CONDUCTOR.

CONDUCTIVITY (Thermal) – Conductivity of heat. The ability of a substance to conduct heat.

CONDUCTOR (Electrical) – Substance which allows electricity to pass through it. Metals are good conductors. See INSULATOR.

CONDUCTOR (Thermal) – Substance which allows heat to pass through it by CONDUCTION.

CONVECTION – The process by which HEAT transfers through a LIQUID or a GAS by CONVECTION CURRENTS. See VENTILATION.

CONVECTION CURRENTS – Movement in LIQUIDS and GASES when heated. Heated particles expand, become less dense and rise. Their place is taken by colder particles.

CO-POLYMER – POLYMER produced from a combination of more than one kind of MONOMER.

CORROSION – CHEMICAL REACTION on the surface, especially of metals which corrode by the action of water, air and chemicals. See GALVANIC CORROSION; RUST; TARNISH; VERDIGRIS.

CRYPTOEFFLORESCENCE – The EFFLORESCENCE of very minute CRYSTALS, i.e. cryptocrystals.

CRYSTALLINE – Composed of CRYSTALS when in SOLID form. See AMORPHOUS.

CRYSTALLIZATION – Process during which CRYSTALS form and link up, as in the solidification of molten metals, and the HYDRATION of plaster and cement.

CRYSTALS – Particles of regular, definite geometric shape which form CRYSTALLINE substances. Many SOLID substances, when pure, are crystalline. See CRYSTALLIZATION.

CURED – See CURING.

CURING – CHANGE OF STATE from LIQUID to SOLID by a CHEMICAL REACTION. General term for the hardening process of plastics, plasters and cements. See ACCELERATOR; HYDRATION; POLYMERIZATION.

DECIBEL (dB) – Unit used to measure levels of intensity of sound. The dB(A) scale takes into account the FREQUENCY as well as the VOLUME of sound for establishing noise levels.

DEHUMIDIFIER – (1) An appliance which draws in humid air, removes the MOISTURE VAPOUR from it, and expels dry air. (2) A HYGROSCOPIC SALT which absorbs MOISTURE VAPOUR from the air to reduce HUMIDITY.

DEHYDRATION – Elimination or removal of water. Usually applies to the removal of chemically combined water, e.g. WATER OF CRYSTALLIZATION.

146

DELIQUESCENT – Having the property to form a solution with MOISTURE VAPOUR in the air. See HYGROSCOPIC.

DENSITY – The mass (weight) of a unit volume of a substance. Expressed in kg/m³ or g/cc. Numerically equal to SPECIFIC GRAVITY.

DEW POINT – The TEMPERATURE at which MOISTURE VAPOUR in the air saturates the air and CONDENSATION takes place. See SATURATED VAPOUR.

DIFFERENTIAL EXPANSION – Many substances expand at different rates when heated. When dissimilar materials are joined together and heated, the differential rates of expansion cause them to either separate or distort. See BI-METAL STRIP.

DILUENT – LIQUID added to SOLUTIONS in order to DILUTE or reduce the VISCOSITY.

DILUTE – (1) To add more SOLVENT or DILUENT to a LIQUID in order to reduce its VISCOSITY. (2) To alter the SOLUTE/SOLVENT ratio of a SOLUTION to make it weaker.

DISPERSION – See COLLOIDAL SOLUTION.

DRY ROT – Various FUNGI which feed on, and destroy wood by reducing it to a dry, weak state. The most common dry rot fungi are: True Dry Rot (*Merulius Lacrymans*), Cellar Fungus (*Poria Vaillantii*). See WET ROT; FUNGICIDE.

ECHO – Effect produced when sound reflects from a solid obstacle. See REVERBERATION.

EFFLORESCENCE – (1) The property of some CRYSTALLINE SALTS to lose some of their WATER OF CRYSTALLIZATION and become powdery on the surface. (2) A white CRYSTALLINE deposit on new cement caused by SALTS in SOLUTION rising to the surface and CRYSTALLIZING as the water is lost by EVAPORATION into the air.

EFFORT – Force applied to move a LOAD, or to bring about a state of EQUILIBRIUM.

ELASTICITY – The property of a substance which has been squashed, stretched or twisted to resume its original shape and size when forces acting upon it are removed.

ELECTRIC CURRENT – A flow of ELECTRONS through a CONDUCTOR. See STATIC ELECTRICITY.

ELECTRO-CHEMICAL – Any chemical process making use of ELECTROLYSIS. See ELECTRO-CHEMICAL SERIES.

ELECTROCHEMICAL SERIES (Electromotive Series) – A series of metals arranged in such an order that a metal which has a high position can displace a lower metal from its SALT, e.g. iron displaces metallic copper from copper sulphate. See GALVANIC SERIES.

ELECTRODE – CONDUCTOR by which an ELECTRIC CURRENT enters or leaves an ELECTROLYTE during ELECTROLYSIS. See ANODE; CATHODE.

ELECTROLYSIS – The chemical decomposition of a liquid by an ELECTRIC CURRENT. See ELECTROLYTE; ANODE; CATHODE.

ELECTROLYTE – Molten solid or a SOLUTION that CONDUCTS an ELECTRIC CURRENT. See ELECTROLYSIS; GALVANIC ACTION; PRIMARY CELL.

ELECTROLYTIC CORROSION – CORROSION caused by ELECTROLYSIS. Produced when two different metals are coupled together in the presence of an ELECTROLYTE. A SIMPLE CELL is formed. See GALVANIC CORROSION.

ELECTROMAGNETIC ENERGY - ENERGY which has wave-like properties. There is a wide range of varying FREQUENCIES and wave lengths, e.g. gamma-rays, X-rays, ultraviolet rays, visible light rays, infra-red (heat) rays, and radio waves. See RADIATION; REFLECTANCE; REFLECTANCE VALUE; REVERBERATION.

ELECTRON – Elementary particles which orbit the nucleus of ATOMS, and bear a charge of NEGATIVE ELECTRICITY.

ELECTRO-PLATING – Depositing a layer of metal on the surface of another metal by ELECTROLYSIS. See ANODIZING.

ELEMENT – Substance consisting entirely of the same ATOMS. Basis of all known substances. See

EMULSION – Minute droplets of LIQUID dispersed in another liquid. See SOLUTION; SUSPENSION.

ENERGY – Capacity for doing work. Various forms of energy, all inter-convertible, include – HEAT, electrical, chemical, radiant, potential and KINETIC energy. See TEMPERATURE; WATT.

EQUILIBRIUM – A state of balance between opposing forces, i.e. when the forces exerted by a LOAD and an EFFORT are balanced.

EVAPORATE – LIQUIDS evaporate at the surface by losing MOLECULES into the air by EVAPORATION. Change from LIQUID to GAS.

EVAPORATION – CHANGE OF STATE from LIQUID to GAS at a TEMPERATURE below the boiling point. See EVAPORATE; HUMIDITY; SATURATED ATMO-SPHERE.

EVAPORATION RATE – The speed at which LIQUIDS, usually DILUENTS and SOLVENTS, EVAPORATE into the atmosphere.

EXOTHERMIC HEAT – Heat given out. ENERGY released in the form of HEAT during a CHEMICAL REACTION. i.e. during HYDRATION or POLYMER-IZATION.

EXOTHERMIC PROCESS – CHEMICAL REACTION or process in which ENERGY in the form of EXOTHERMIC HEAT is released.

FATTY ACID – ORGANIC ACID, many of which occur in living things, usually in fats and oils. See SAPONIFICATION.

FERROUS METAL – Metal containing iron, e.g. iron and steels. See NON-FERROUS METAL.

FILTER – Device used to separate. 1. COLOUR filters separate WHITE LIGHT by allowing only certain wavelengths to pass through. See ELECTRO-MAGNETIC WAVES; SPECTRUM. 2. Other filters separate SOLIDS from SUSPENSIONS by allowing only LIQUIDS and dissolved substances to pass through.

FLAMMABLE – Easily set on fire. Usually related to LIQUIDS and GASES which will burn. See COMBUSTIBLE; FLASH POINT; HEAT OF COMBUSTION.

FLASH POINT – The lowest TEMPERATURE at which a substance gives off sufficient FLAMMABLE VAPOUR to ignite when a small flame is applied. See IGNITION POINT.

FLEXIBILITY – The property of a substance which allows it to be distorted by bending and stretching without breaking.

FLUORESCENCE – Ability of some substances to ABSORB ENERGY of one wavelength, and emit visible light of another wavelength. See ELECTROMAGNETIC WAVES; FLUORESCENT.

FLUORESCENT – A substance is fluorescent if it exhibits FLUORESCENCE, e.g. fluorescent lamps contain powders which emit LIGHT when an electrical discharge is passed through a gas in the lamp.

FREQUENCY – In wave motions, the number of vibrations or waves per second. See DECIBEL; RADIATION; VOLUME; WAVELENGTH.

FRICTION – Forces which offer resistance to movement between surfaces in contact with each other. See LUBRICANT.

FUNGI – Plural for FUNGUS.

FUNGICIDAL SOLUTION – Water or spirit based SOLUTION containing TOXIC ingredients kill FUNGI and MOULDS. See DRY ROT; STERILIZE.

FUNGICIDE – Any preparation which kills FUNGI and MOULDS.

FUNGUS – Simple parasitic plant life which feeds on ORGANIC matter. Ranges from microscopic MOULDS to mushrooms and toadstools. See DRY ROT.

GALVANIC ACTION – CHEMICAL REACTION between different metals when coupled together in an ELECTROLYTE. See ELECTROLYSIS; PRIMARY CELL; SACRIFICIAL PROTECTION.

GALVANIC CORROSION – The corrosive effect upon metals due to ELECTROLYSIS. See GALVANIC ACTION.

GALVANIC PROTECTION – See SACRIFICIAL PROTECTION.

GALVANIC SERIES – A re-arrangement of the ELECTROCHEMICAL SERIES of the metals for the study of CORROSION.

GALVANIZING – The process of coating iron and steel with zinc by dipping into molten zinc, or by ELECTRO-PLATING. When solidified, the layer of zinc offers SACRIFICIAL PROTECTION to the iron. See SHERARDIZING.

GAS – A STATE OF MATTER in which the MOLECULES are spread very far apart, and move around the whole of the space in which the gas is contained. See EVAPORATION; VAPOUR.

GEL – COLLOIDAL SUSPENSION which forms a jelly, the VISCOSITY being so great that it has the ELASTICITY of a SOLID.

GRAVITY – Attraction existing between all matter, i.e. the Earth's attraction to everything on it and surrounding it. See ATMOSPHERIC PRESSURE; DENSITY.

HARDEN – To become SOLID, stronger, resistant to crushing and scratching, e.g. plasters and cements SET quite soon after mixing and then harden, or CURE at a much slower rate. See HYDRATION; REVERSIBLE COATING.

HARDENER – See CATALYST; CURING.

HEAT – ENERGY contained in matter in the form of the KINETIC ENERGY of its MOLECULES.

HEAT CAPACITY – The amount of HEAT a substance will ABSORB. The amount of heat needed to raise the TEMPERATURE of a body 1 degree CELSIUS.

HEAT OF COMBUSTION (Calorific Value) – The amount of HEAT ENERGY given out from a certain quantity of a substance when burned in oxygen. See CALORIE; INTERNAL COMBUSTION.

HEAT OF HYDRATION – HEAT ENERGY released during the HYDRATION of plasters and cements. See EXOTHERMIC HEAT.

HUMIDITY – A measure of the MOISTURE VAPOUR present in the air. See RELATIVE HUMIDITY; EVAPORATION; SATURATED ATMOSPHERE.

HYDRATE – COMPOUND containing chemically combined water in the form of WATER OF CRYSTALLIZATION. See HYDRATION.

HYDRATION – CHEMICAL REACTION of a substance combining with water. Generally applied to combining with WATER OF CRYSTALLIZATION as in the hardening process of plasters and cements.

HYDRAULIC CEMENT – Cement which will HARDEN in contact with water without air being present. See CRYSTALLIZATION; HYDRATION.

HYDRAULIC PRESSURE – Fluid pressure. The pressure applied anywhere to an enclosed VOLUME of LIQUID is transmitted equally, in all directions, and acts on every portion of the container. See ATOMIZATION; HYDRAULICS.

HYDRAULICS – Practical application of the study of movement of LIQUIDS to convey power.

HYDROLYSIS – The chemical reaction of a substance with water. See SAPONIFICATION.

HYDROMETER – Instrument for measuring the RELATIVE DENSITY or SPECIFIC GRAVITY of LIQUIDS.

HYGROMETER – Instrument for measuring the RELATIVE HUMIDITY of the atmosphere.

HYGROSCOPIC – Having a tendency to ABSORB MOISTURE VAPOUR from the air. Many SALTS are hygroscopic. See DEHUMIDIFIER; DELIQUESCENT.

IGNITE – To set on fire. To introduce a flame, or other course of HEAT, to bring about COMBUSTION. See FLASH POINT; IGNITION POINT.

IGNITION POINT – TEMPERATURE to which a substance must be heated before it can be IGNITED and COMBUSTION take place. See FLASH POINT.

IGNITION TEMPERATURE – See IGNITION POINT.

INDICATOR – Substance which changes colour when in contact with SOLUTIONS of various pH VALUES. Used to indicate degree of acidity or alkalinity. See ACIDIC; NEUTRALIZATION; UNIVERSAL INDICATOR.

INERTIA – The tendency of a body to preserve its state of rest if at rest, and to continue moving if moving.

INFRA-RED RADIATION – RADIATION of invisible HEAT rays from a source such as the Sun or a radiant fire. ELECTROMAGNETIC WAVES just beyond the red end of the SPECTRUM, with energy between that of visible light and radio waves (microwaves). See RADIANT HEAT.

INITIAL SET – General term for the initial stage in changing from a LIQUID to a SOLID. The substance becomes less plastic and increases VISCOSITY, e.g. plaster, paints and plastics go through initial set before HARDENING.

INSULATION – (1) Means of reducing the passage of electricity, heat, or sound by the use of an INSULATOR. (2) Preventing the passage of moisture through ABSORBENT materials by the use of a non-absorbent membrane. See CONDUCTOR.

INSULATOR – (1) Substance which is a non-CONDUCTOR of electricity. E.g. rubber or plastic. (2) Substance which is reluctant to allow sound, heat, light or moisture to pass through it.

INTERNAL COMBUSTION – The COMBUSTION of fuel and air above a piston inside an enclosed cylinder. The ENERGY released is converted to mechanical movement by the piston. Principle of petrol and diesel internal combustion engines (I.C.E.). See HEAT OF COMBUSTION.

INTUMESCENT – Intumescent materials are those which swell-up, or foam when heated (intumesce). E.g. some types of fire retardant coatings.

JOULE (J) – Unit of work or ENERGY. 1 CALORIE = 4.2 J; 1 J = 0.24 calories.

KELVIN (Absolute Scale) (°K) – Unit of thermodynamic TEMPERATURE. Scale ranging upwards from ABSOLUTE ZERO, with identical intervals as the CELSIUS scale.

KILOJOULE (kJ) – 1000 JOULES. Unit of work or ENERGY. 1 kJ = 240 CALORIES.

KINETIC ENERGY – The ENERGY which a body possesses by virtue of its motion.

LATENT HEAT – Heat ABSORBED or released without a change in TEMPERATURE when a substance changes its STATE OF MATTER.

LIGHT - ELECTROMAGNETIC WAVES which can be seen by the eye. Variations in WAVELENGTHS produce different sensations of COLOUR. See REFLECTANCE VALUE.

LIQUEFIED – A substance is liquefied when it is changed into a LIQUID from another STATE OF MATTER, e.g. CONDENSATION, LIQUEFIED GAS, melting.

LIQUEFIED GAS – GAS converted into a LIQUID state by lowering its TEMPERATURE and/or increasing its pressure and maintaining it in a liquid state by storing in a pressurized container. E.g. butane, propane. See CHANGE OF STATE; REFRIGERANT.

LIQUID – One of the STATES OF MATTER intermediate between a SOLID and a GAS. Liquids flow and assume the shape of their container.

LITMUS – Purple substance of vegetable origin used as an INDICATOR. Turned red by ACIDS, and blue by ALKALI. See UNIVERSAL INDICATOR.

LOAD – (1) Mechanical force or weight applied to a body. (2) The weight supported by a structure. See EFFORT; EQUILIBRIUM.

LUBRICANT – Substance used to reduce the FRICTION between SOLIDS moving against one another. E.g. oil.

LUMEN – Unit of luminous flux, i.e. the amount of light emitted per second in a cone of a certain size by a point source of known intensity.

LUX – Unit of illumination. One LUMEN per square metre.

MAGNETIC FIELD – An area of lines of magnetic force which exist around a magnet, or a coil of wire carrying an ELECTRIC CURRENT. See MAGNETIC POLE.

MAGNETIC POLE – A magnet appears to have its lines of force concentrated at two points, near the ends, the pole. If suspended and allowed to swing, the N pole will point North.

MECHANICAL ADHESION – ADHESION through the mechanical interlocking of one substance into the surface irregularities of another, usually a SOLID. See SPECIFIC ADHESION.

MECHANICAL ADVANTAGE (MA) – In a machine the ratio of the LOAD moved to the EFFORT needed to maintain the movement. See EQUILIBRIUM.

MEGAJOULE (MJ) – Unit of work or ENERGY. One million JOULES. 1 MJ = 240 000 CALORIES.

MIXTURE – Two or more substances blended together with *no* chemical bonding between them. Each substance retains its individual properties and the mixture can be separated by physical means. See SOLUTION.

MOISTURE CONTENT (MC) – The amount of moisture contained in the air spaces of POROUS materials. Expressed as a percentage of the total amount of air spaces in the material sample. See ABSORBENCY; CAPILLARY ACTION; MOISTURE METER.

MOISTURE METER – Instrument used to obtain a quick reading of MOISTURE CONTENT.

MOISTURE VAPOUR – Water vapour. Water in a gaseous STATE OF MATTER. EVAPORATION of water into the air produces HUMIDITY. See CONDENSATION; DELIQUESCENT; HYGROMETER.

MOLECULE – Two or more ATOMS joined together. The smallest part of a compound capable of existing independently while retaining the properties of the compound. See CHEMICAL REACTION; POLYMERIZATION.

MONOMER – Simple chemical COMPOUND whose molecules can be linked in long chains to form a POLYMER.

MOULD – Microscopic FUNGI which feed on ORGANIC matter such as paper and fabric. E.g. mildew. See FUNGICIDAL SOLUTION.

NEGATIVE ELECTRICITY – Particles bearing a negative electrical charge, i.e. ELECTRONS. When a substance is charged with electricity by rubbing with another substance, two electrical charges are produced in equal amounts. One substance contains a negative charge, sign — , and the other a positive charge, sign + . Objects bearing the same charge repel each other. Objects bearing different charges attract each other.

NEUTRAL – Neither ACIDIC nor ALKALINE. Having a pH VALUE of 7, i.e. pure water. See NEUTRALIZATION.

NEUTRALIZATION – CHEMICAL REACTION between an ACID and an ALKALI to produce a NEUTRAL SOLUTION OF A SALT.

NON-FERROUS METAL – Metal containing no iron, e.g. copper, zinc and lead. See FERROUS METAL.

ORGANIC – Relating to all carbon compounds except the very simple ones such as carbon dioxide. Organic substances exist as a result of residues from animal and plant life (oil, etc.) or are formed by living organisms, or they can be artificially made. See CARBON CHAIN.

OXIDATION – CHEMICAL REACTION or combining with oxygen. E.g. COMBUSTION, or the production of an OXIDE of a metal. See OXIDIZER; RUST.

OXIDE – Chemical COMPOUND of oxygen and other ELEMENTS. E.g. RUST is an oxide of iron. See OXIDATION.

OXIDIZER – A substance that causes the process of OXIDATION. See BLEACH.

PERMEABILITY – A measure of the extent that a substance is PERMEABLE.

PERMEABLE – A body is permeable to a substance if it allows the passage of the substance through itself.

PHOSPHATING (Parkerizing) – Process of forming a RUST resistant SALT on iron and steel by immersing in a hot SOLUTION of phosphoric ACID and manganese. See NEUTRALIZATION.

PHYSICAL CHANGE – Any CHANGE OF STATE which does not involve a CHEMICAL REACTION. E.g. ice to water.

pH VALUE – Measure of how ACIDIC or ALKALINE a SOLUTION is. Scale of numbers from 1, strongly acidic, to 7, NEUTRAL to 14, strongly alkaline. See INDICATOR; LITMUS.

PLASTICIZER – Substance added to SOLIDS to improve the properties of ELASTICITY and FLEXIBILITY.

POLYMER – Large MOLECULE consisting of chemically joined MONOMERS. Product of POLYMERIZATION. See CONDENSATION POLYMER.

POLYMERIZATION – Chemical union of MONOMERS to form larger MOLECULES called POLYMERS which link-up (cross link) with each other to form a new substance. See ADDITION POLYMER; CHEMICAL REACTION.

POROSITY – General term for the amount of air spaces in POROUS substances.

POROUS – Porous substances are those which have air spaces in their composition and possess POROSITY.

PRECIPITATE – An insoluble substance formed in a SOLUTION as a result of a CHEMICAL REACTION.

PRECIPITATION – The process of making a PRECIPITATE.

PRIMARY CELL (Simple Cell) – Device for producing an ELECTRIC CURRENT by a CHEMICAL REACTION. Usually consists of two different metals coupled together in an ELECTROLYTE. See GALVANIC ACTION.

RADIANT HEAT – See INFRA-RED RADIATION.

RADIATION – General term for the emission of rays, wave motions, or particles from a source. Usually applies to ELECTROMAGNETIC WAVES. See INFRA RED RADIATION; REFLECTANCE.

REFLECTANCE – A measure of the amount of RADIATION a surface is capable of reflecting.

REFLECTANCE VALUE (Light Reflectance Value) (L.R.V.) – A measure of the amount of LIGHT a coloured surface will reflect. Expressed as a percentage of the total light falling on it, e.g. a white surface reflects about 98%.

REFLECTIVE SURFACE – Surface which reflects RADIATION falling upon it. Smooth surfaces with a high REFLECTANCE VALUE are efficient reflectors of LIGHT and HEAT. See ECHO; REVERBERATION.

REFRIGERANT – LIQUID used to transfer HEAT from one place to another in refrigeration. Usually a LIQUEFIED GAS. See CONDUCTION.

RELATIVE DENSITY – See SPECIFIC GRAVITY.

RELATIVE HUMIDITY (RH) – A measure of HUMIDITY. The amount of MOISTURE VAPOUR present in the air, compared to the amount needed to SATURATE the air at the same TEMPERATURE. Shown as a percentage and measured with a HYGROMETER.

RETARDER – Substance which slows the rate of a CHEMICAL REACTION. See ACCELERATOR.

REVERBERATION – Repeated reflection of RADIATION or wave motions from a surface, e.g. ECHO is the reverberation of sound waves.

REVERSIBLE COATING – Surface coating (paint etc.) which HARDENS by a PHYSICAL CHANGE brought about by the EVAPORATION of the SOLVENT. See SOLUTION.

REVERSIBLE FILM – See REVERSIBLE COATING.

RUST – Hydrated OXIDE of iron. Forms on the surface of iron when exposed to air and moisture. See CORROSION.

RUST INHIBITIVE – Property of some pigments, when used in paints, to reduce the spread of RUST beneath the paint film.

SACRIFICIAL COATING – A metallic coating, usually zinc, applied to iron and steel to give SACRIFICIAL PROTECTION. See CATHODE.

SACRIFICIAL PROTECTION (Galvanic or Cathodic Protection) – The deliberate sacrifice of one metal by GALVANIC CORROSION in order to protect the other metal coupled to it. Principle of GALVANIZING.

SALT – COMPOUND produced by the NEUTRALIZATION of an ACID with an ALKALI. SALTS normally are a compound of a metal. The neutralization of a FATTY ACID produces a SOAP.

SAPONIFICATION – Process of chemical combination of fats with an ALKALI to produce a SOAP.

SAPWOOD – Part of a tree directly below the bark, through which flows the sap to nourish the tree.

SATURATED ATMOSPHERE – Water enters the atmosphere by EVAPORATION until such time as no more molecules of water can be ABSORBED. The air becomes saturated with MOISTURE VAPOUR. Any excess of vapour in the air returns to a LIQUID state by CONDENSATION.

SATURATED SOLUTION – A SOLUTION which can ABSORB no more SOLUTE. Any excess of solute remains undissolved in the SOLVENT.

SATURATED VAPOUR – A VAPOUR which can absorb no more liquid by EVAPORATION until some of the vapour becomes liquid by CONDENSATION.

SET – See INITIAL SET.

SHEAR – To cut or slice. To distort or break by a shearing force such as twisting.

SHERARDIZING – Similar to GALVANIZING except that the iron is heated together with zinc dust.

SIMPLE CELL – See PRIMARY CELL.

SOAP – Compound produced by the SAPONIFICATION of a fat. See NEUTRALIZATION; SALT.

SOLID – STATE OF MATTER in which the substance retains its shape by strong forces of COHESION between its MOLECULES.

SOLUBILITY – The extent to which a SOLUTE will dissolve in a SOLVENT to form a SOLUTION.

SOLUTE – Substance which is dissolved into a SOLVENT to form a SOLUTION. See SOLUBILITY; SOLVENT POWER.

SOLUTION – Two or more substances mixed together so that their MOLECULES are completely intermingled. Most commonly SOLIDS dissolved in LIQUIDS. See SOLUTE; SOLVENT; COLLOIDAL SOLUTION.

SOLVENT – (1) LIQUID having the power to dissolve other substances into itself to form a SOLUTION. (2) Part of a solution which has the same STATE OF MATTER as the solution itself. See SOLVENT POWER.

SOLVENT POWER – The extent to which a SOLVENT disperses the MOLECULES of a substance dissolved into it when forming a SOLUTION.

SPECIFIC ADHESION – ADHESION of one substance to another by the attraction of MOLECULES. See MECHANICAL ADHESION.

SPECIFIC GRAVITY (SG) (Relative Density) – The ratio of the DENSITY of a substance to the density of water. Numerically equal to the density in g/cc.

SPECIFIC HEAT CAPACITY – See HEAT CAPACITY.

SPECTRAL RANGE – Colour range of the SPECTRUM.

SPECTRUM – Arrangement of the different kinds of ELECTROMAGNETIC ENERGY in order of WAVELENGTHS. With respect to light, the colours in order of decreasing wavelengths are: red, orange, yellow, green, blue, indigo, violet. See ELECTROMAGNETIC ENERGY.

SPONTANEOUS COMBUSTION – COMBUSTION without an external source of IGNITION. Combustion results from the heat produced within the substance by slow OXIDATION. See EXOTHERMIC HEAT.

SPORE – The reproductive cell of certain plants such as moss, FUNGI and ferns. When set free under suitable conditions gives rise to a new plant. See MOULDS; FUNGICIDE.

STATES OF MATTER – The three physical states in which matter exists, i.e. SOLID, LIQUID or GAS. See CHANGE OF STATE.

STATIC ELECTRICITY – Electricity at rest, as

opposed to an ELECTRIC CURRENT. Produced by 'charging' a body with electricity by FRICTION from rubbing substances together.

STERILIZE – To free a surface of all bacteria and MOULD SPORES by treatment with HEAT or a STERILIZING SOLUTION. See FUNGICIDE.

STERILIZING SOLUTION – LIQUID which will STERILIZE a surface to which it is applied. See FUNGICIDE; TOXIC WASH.

SUCTION – A LIQUID raised up by suction is actually pushed up by outside ATMOSPHERIC PRESSURE, which is greater than the pressure inside which has been made lower by increasing the VOLUME.

SURFACE TENSION – Tendency of the surface of a LIQUID to contract and show properties similar to that of a stretched elastic film. Caused by forces between the molecules.

SUSPENSION – Very small SOLID particles distributed in a LIQUID. See COLLOIDAL SOLUTION.

TARNISH – Surface discoloration on some NON-FERROUS METALS, usually due to a film of a COMPOUND of the metal. See VERDIGRIS.

TEMPERATURE – A measure of the KINETIC ENERGY of ATOMS or MOLECULES in a substance. The degree of 'hotness' or 'coldness' of a substance. Measured in degrees with a THERMOMETER. See ABSOLUTE ZERO; CELSIUS; KELVIN.

THERMAL CONDUCTIVITY – See CONDUCTIVITY (Thermal).

THERMOMETER – Instrument for measuring TEMPERATURE. Calibrated in degrees.

THERMOPLASTIC – Becomes plastic (softens) on being heated, and HARDENS on cooling, the process being completed as many times as necessary. See THERMOSETTING; PHYSICAL CHANGE.

THERMOSETTING – Not THERMOPLASTIC. Having once undergone a CHEMICAL REACTION by heating, is resistant to further moulding and shaping by heat treatment.

THERMOSTAT – Automatic device for maintaining an even TEMPERATURE. It cuts-off the supply of HEAT when the required temperature is reached, and restores it as the temperature falls. See BI-METAL STRIP.

THIXOTROPIC – Substance possessing the properties of THIXOTROPY.

THIXOTROPY – The property of some LIQUIDS, i.e. paints, to increase in VISCOSITY if left undisturbed, but on shaking or stirring the viscosity reduces to its original value.

TOXIC – Poisonous. See FUNGICIDAL SOLUTION.

TOXIC WASH – Poisonous LIQUID used to STERILIZE surfaces on which are MOULD SPORES and bacteria.

TRANSLUCENT – Semi TRANSPARENT. Allows partial transmission of LIGHT. Permits light to pass through in such a way that an object cannot be clearly seen through it, e.g. frosted glass.

TRANSPARENT – Allows all LIGHT to be transmitted. Permits light to pass through in such a way that an object can be clearly seen through it. E.g. clear glass. See TRANSLUCENT.

ULTRAVIOLET – RADIATION of invisible light just beyond the violet end of the SPECTRUM. ELECTROMAGNETIC WAVES with ENERGY between that of visible light and X-rays. See INFRA-RED RADIATION.

UNIVERSAL INDICATOR – Mixture of INDICATORS selected to give a gradual colour change over a range of pH VALUES from 1 to 14. See LITMUS.

VACUUM – Space in which there are no ATOMS or MOLECULES. Generally taken to mean a space containing air, or other GAS at a lower pressure than ATMOSPHERIC PRESSURE.

VAPOUR – Substance in a gaseous STATE OF MATTER, which may be LIQUEFIED by increasing the pressure. See CONDENSATION; HUMIDITY; LIQUEFIED GAS; SATURATED VAPOUR.

VENTILATION – Means of changing a SATURATED ATMOSPHERE or stale air for fresh, dry air.

Achieved with CONVECTED CURRENTS, or draughts induced by fans or blowers.

VENTURI – A tapered throat inside a tube which accelerates a stream of air passing through it.

VERDIGRIS – Green coating formed upon copper due to a basic SALT of the metal forming in a HUMID atmosphere. See TARNISH.

VISCOMETER – Instrument for measuring the VISCOSITY of LIQUIDS.

VISCOSITY – Tendency of a LIQUID to resist motion within itself. Resistance to flow. Measured with a VISCOMETER.

VOLT (V) – Unit of electrical 'force' or 'pressure' which produces an ELECTRIC CURRENT. See AMPERE; WATT.

VOLTAGE – The pressure of an ELECTRIC CURRENT through a CONDUCTOR, measured in VOLTS.

VOLUME – (1) General term to indicate the loudness or intensity of sound. See DECIBEL; FREQUENCY. (2) A measure of space occupied by a substance.

WATER OF CRYSTALLIZATION – MOLECULES of water chemically combined with the molecules of certain CRYSTALLINE substances. See HYDRATE.

WATER VAPOUR – See MOISTURE VAPOUR.

WATT (W) – Unit of power. One JOULE per second. The ENERGY of an ELECTRIC CURRENT. Electrical appliances are rated by their consumption of energy in watts. $W = AMPERES \times VOLTS$.

WAVELENGTH – Distance between successive points of a wave motion, i.e. the distance from crest to crest. See ELECTROMAGNETIC WAVES; RADIATION.

WET ROT – Various FUNGI which feed on and destroy wood which is maintained in a very wet condition (50 to 60% MOISTURE CONTENT). The most common wet rot fungus is called cellar fungus (*Coniophora cerebella*). See DRY ROT.

WOODWORM – General term for the larva, or grub, of certain beetles which infest wood. Common types include furniture beetle (*Anobium punctatum*), death watch beetle (*Xestobium rufovillosum*) and powder post beetle (*Lyctus*).

INDEX

Terms shown in capital letters in the text are defined in the Glossary. This Index makes no reference to the Glossary.

157

162